SUPER POLLUTERS

SOCIETY AND THE ENVIRONMENT

The impact of humans on the natural environment is one of the most pressing issues of the twenty-first century. Key topics of concern include mounting natural resource pressures, accelerating environmental degradation, and the rising frequency and intensity of disasters. Governmental and non-governmental actors have responded to these challenges through increasing environmental action and advocacy, expanding the scope of environmental policy and governance, and encouraging the development of the so-called "green economy." Society and the Environment encompasses a range of social science research, aiming to unify perspectives and advance scholarship. Books in the series focus on cutting-edge global issues at the nexus of society and the environment.

SUPER POLLUTERS

TACKLING THE WORLD'S LARGEST SITES OF CLIMATE-DISRUPTING EMISSIONS

DON GRANT,
ANDREW JORGENSON,
AND
WESLEY LONGHOFER

Columbia University Press *New York*

Columbia University Press
Publishers Since 1893
New York   Chichester, West Sussex
cup.columbia.edu

Library of Congress Cataloging-in-Publication Data
Names: Grant, Don S., author. | Jorgenson, Andrew, author. | Longhofer, Wesley, author.
Title: Super polluters: tackling the world's largest sites of climate-disrupting emissions / Don Grant, University of Colorado, Andrew Jorgenson, Boston College, Wesley Longhofer, Emory University.
Description: New York: Columbia University Press, [2020] | Includes bibliographical references and index.
Identifiers: LCCN 2020012997 (print) | LCCN 2020012998 (ebook) | ISBN 9780231192163 (hardback) | ISBN 9780231192170 (paperback) | ISBN 9780231549691 (ebook)
Subjects: LCSH: Electric power-plants—Environmental aspects. | Air—Pollution. | Carbon dioxide mitigation. | Environmental sociology.
Classification: LCC TD195.E4 G73 2020 (print) | LCC TD195.E4 (ebook) | DDC 363.738/741–dc23
LC record available at https://lccn.loc.gov/2020012997
LC ebook record available at https://lccn.loc.gov/2020012998

Columbia University Press books are printed on permanent and durable acid-free paper.
Printed in the United States of America

Cover design: Julia Kushnirsky
Cover illustration: ©istockphoto

This book is dedicated to Natasha, Allison, Kelsey, Roger, Penelope, Harper, and Juniper: part of the next generation who will inherit our planet and zap up its energy in the best ways possible.

CONTENTS

ILLUSTRATIONS AND TABLES

Illustrations

Tables

ACKNOWLEDGMENTS

WE would like to thank Casey Gittelman, Julia Mulliez, Urooj Raja, and Jamie Vickery for their research assistance and Kelly Bergstrand, Katrina Running, Bodi Vasi, and Richard York for their collaboration on published research projects that we build on here. We also thank Sarah Babb, Brett Clark, Thomas Dietz, Liam Downey, Jeffrey Logan, Doug McAdam, Juliet Schor, and all the members of the Environmental Sociology Working Group at Boston College for comments on individual chapters and the research for this book as it evolved. For additional feedback on the research presented in this book, we thank the attendees of invited talks at the Department of Sociology, University of California, Riverside; Department of Sociology, University of California, Irvine; Department of Sociology, University of British Colombia; Department of Sociology, Michigan State University; Ecology Center, Utah State University; Department of Sociology, Vilnius University; Department of Resource Economics and Environmental Sociology, University of Alberta; Department of Civil and Environmental Engineering, Tufts University; Department of Natural Resources, Cornell University; Transnational Studies Initiative and Department of Sociology, Harvard University;

Gordon Research Conference for Industrial Ecology; National Renewable Energy Laboratory; Social Science Environmental Health Research Institute, Northeastern University; School of Forestry and Environmental Studies, Yale University; Department of Earth and Environmental Sciences, Boston College; Law School, Emory University; and School of Public Policy, Georgia Institute of Technology.

We would like to thank Dana Fischer, Lori Peek, and Evan Schofer, our book series editors, and Eric Schwartz and Lowell Frye at Columbia University Press for their guidance, support, and encouragement. We also thank anonymous reviewers for their extremely thoughtful and constructive commentary.

Funding for this project was provided by the Sociology Program of National Science Foundation (Grant Numbers 1123262, 1357483, 1357495, and 1357497). Previous publications by the authors are the bases of chapters 2 (Grant, Jorgenson, and Longhofer 2013; Jorgenson, Longhofer, and Grant 2016), 3 (Grant, Jorgenson, and Longhofer 2018), 4 (Grant, Running, Bergstrand, and York 2014; Grant, Jorgenson, and Longhofer 2016), and 5 (Grant, Bergstrand, and Running 2014; Grant and Vasi 2017).

SUPER POLLUTERS

1

WHO IS RESPONSIBLE FOR THIS MESS?

The Climate Crisis and Hyperemitting Power Plants

> One problem that climate change activists have had in trying to mobilize action is the difficulty of . . . identifying specific villains to blame for the escalating threat. Instead, the crisis seems to be largely the product of impersonal forces beyond our control.
>
> Doug McAdam (2017)

TAIWAN'S Taichung power plant looms above the island's central western coastline. Spread over three square miles and having a productive capacity of nearly 6,000 megawatts, the massive state-owned and -operated electrical generating station is the largest thermal power plant ever constructed. It transmits current across miles of cable to the country's rapidly growing core, where it energizes upscale neighborhoods, state-of-the-art schools, boutique hotels, high-tech factories, and spectacular cultural and recreational venues. Providing over 20 percent of its nation's total electrical output, the Taichung plant has been instrumental in transforming Taiwan from one of the poorest economies into one of the richest.

Unfortunately, this engine of prosperity is also the world's biggest emitter of carbon dioxide (CO_2). Its seventy-five-story smokestacks, each adorned with painted white birds, spew over

FIGURE 1.1 Taichung Power Plant, Longjing District, Taichung City, Taiwan.
Source: Chongkian, via the Creative Commons Attribution-ShareAlike 3.0 Unported License, https://commons.wikimedia.org/wiki/File:Taichung_Thermal_Power_Plant.JPG

40 million tons of carbon into the atmosphere each year, roughly equivalent to the amount released by Switzerland. If you were to visit the power station, though, you might get the impression it does not pollute at all. Proud of the engineering feat this largest thermal power plant represents, the state conducts daily tours of the facility and even advertises it as a travel destination. Inside the "giga" project, sightseers are shown numerous exhibits, including the architects' original miniature model of the complex and a display explaining the science behind its turbine engines. They are also treated to a viewing of the enormous mechanical system itself. But nary a word or sign about the egregious volume of carbon the plant emits. Likewise, when led outside beneath the towering stacks, visitors will find it hard to detect anything

menacing flowing from them. Thanks to advanced combustion systems, black smoke never leaves the chimneys as it does at older plants. And in the summer, because the waste air rarely condenses into water vapor, they will likely see no smoke of any sort.

Still, there is no mistaking the 14.5 million tons of coal that are annually delivered to the station to be burned. This and skyrocketing death rates from lung-related cancer since the plant was commissioned in the early 1990s have raised concerns about its environmental and health impacts. When confronted by reporters and activists about the station's emissions, spokesperson and chief engineer Tu Yueh-yuan downplayed their significance, claiming the plant emits more pollutants than others because it services a larger share of the country's electricity needs. She added that management is committed to reducing discharges and, toward that end, is considering plans to upgrade the efficiency of the plant's generators.[1]

Power plants are both the lifeblood and the bane of modern society. Electrification has revolutionized media and communication, dramatically improved medical care, spurred the rise of the metropolis, and enabled the global human population to grow by over 6 billion during the past century. But because fossil-fueled power plants are the single largest emitters of anthropogenic greenhouse gases (GHGs), especially CO_2, they also pose one of the greatest threats to humans' life-support system. The International Energy Agency[2] estimates that to have a reasonable (greater than 66 percent) chance of preventing damage to that system, global energy-related carbon emissions must fall by more than 70 percent over the next thirty-five years. Whether the world community possesses the requisite will and wisdom to rein in the worst of its hyperemitting power plants, therefore, could decide its destiny.

Part of the challenge is simply raising awareness about these profligate polluters. Not only do these polluters go to great lengths, like the Taichung plant, to normalize and obfuscate their exorbitant emissions, but also society at large often discusses the topic of climate change in ways that divert attention from them. When asked who is to blame for the record heat waves, wildfires, and destruction caused by hurricanes over the past decade, citizens and commentators will frequently cite individuals (e.g., the Koch Brothers), political parties (e.g., Republicans), or demographic groups (e.g., baby boomers) as the culprits. While these actors all certainly play a role in exacerbating climate change, or at least have a financial or political interest in ignoring it, none is a large-scale polluter per se. Others attribute climatic changes to the major consumers and suppliers of coal, gas, and oil—namely, the rich and the fossil-fuel industry as well as the economic sectors that rely on it, such as transportation and agriculture. Obviously, they, too, are important players and bear significant responsibility for the damage done to the planet. But to focus solely on these purchasers and extractors of energy is to lose sight of the power plants that connect them to the points where energy is actually produced and the largest quantities of carbon are released. Still others, like the International Energy Agency, claim that individual nations are mainly at fault. They stress how developed countries like the United States and rapidly developing ones like China and India have greatly outsized carbon footprints. But that information, by itself, is difficult to act on because it fails to pinpoint where within nations the lion's share of carbon pollution is being emitted.

In response to this shortcoming, many of the same groups have examined which sectors emit the most CO_2 and have found that electricity is the primary culprit, accounting for nearly half of such emissions.[3] However, in focusing on the total emissions of this sector, their reports portray actors inside it as uniformly hostile to ecosystems, when, in fact, some are more environmentally progressive than others. For example, whereas Taichung's power plant continues to be fined for burning more coal than its city permit allows,[4] Xcel's coal-fired plant in Boulder, Colorado, is transitioning to renewable energy sources in response to local protests and that city's attempts to municipalize the facility.[5] It follows that to effectively tackle the problem of climate change, we need to drill down past the sectoral level to distinguish the really bad apples from the merely bad ones.

But sorting out hyperemitting power plants is not enough. We also need to know what makes them discharge more carbon than their counterparts and what can be done about it. The most obvious reason some fossil-fueled power plants have higher emission rates and levels is that they use more carbon-intensive inputs. Indeed, industry and government officials have long argued that the key to transitioning to a green economy is gradually replacing coal and oil with natural gas. While many would agree that fuel switching is a step in the right direction, this approach fails to address not only what the social forces are that lock in dependence on fossil energy[6] but also how those forces may help explain why some plants are especially dangerous compared to others using the same fuel type. For example, the ability and motivation of coal plants to curb emissions may be conditioned by their age, size, and location in the world economy as well as by the civic engagement and policies of their surrounding communities. To suggest that decarbonizing extreme polluters is as simple as substituting one fossil fuel for another is akin to arguing that, even without knowing what factors make criminals prone to violence in the first place, we can reduce crime rates by getting them to use less injurious weapons.

The fact is that our understanding of the social causes and mitigation of the CO_2 emitted by power plants lags far behind our knowledge of those emissions' harmful impacts. The natural sciences have extensively documented the disastrous effects carbon pollution can have and is having with regard to species extinction; the availability of water, food, and shelter; disputes over natural resources; population migration; and adverse human health outcomes. By comparison, social scientific research on the determinants and abatement of electricity-based CO_2 emissions is less developed and, consequently, still occupies a relatively limited space in global climate change studies. It has long been known that human activities such as burning fossil fuels contribute to environmental degradation, but the societal forces behind these activities have largely remained a "black box." Although the social sciences are becoming more integrated into interdisciplinary climate research and international policy development,[7] we argue that this integration has been far too slow and not substantial enough.

A notable exception to the marginalization of the social sciences is the work of several energy and environmental economists that exerts substantial influence over bodies like the Intergovernmental Panel on Climate Change (IPCC). These economists contend that societies can create a sustainable electricity sector by manipulating resource prices and stimulating the development of more efficient technologies. They use sophisticated statistical modeling techniques and deductive (and often highly mathematical) theorizing to generate composite analyses of a nation's or the world's electricity sector. For example, one technique popularized by Nobel Prize–winning William Nordhaus and his colleagues is the integrated assessment model (IAM), which is used to estimate the costs and benefits of global temperature changes, taking into account, in some models, various energy choices and technologies.

IAMs and other methods simulate how the climate and economy evolve together and have been used to answer such questions as how the world could avoid 1.5°C of global warming at the lowest possible cost and what the likely impact is of current country pledges to lower future emissions. Thus, they help researchers to identify which available or planned equipment and devices can optimize resource usage and to reduce the potential causes of and solutions to carbon pollution to a single factor—cost. In turn, researchers can suggest which national energy and climate policies should work best to increase carbon efficiency in a technically optimized free market society.

It is easy to understand why groups like the IPCC that are charged with balancing the demands of science, which requires independence, and those of diplomacy, which requires sensitivity to the preferences of multiple governments,[8] would be attracted to this type of approach. Energy and environmental economists' aggregate-level studies make "wicked problems"[9] like climate change seem more manageable compared, for example, to complex system frameworks that detail the atmosphere's numerous, interdependent, and often unpredictable feedback loops.[10] Focusing on efficiency measures also potentially minimizes conflict between developed and developing nations because optimizing resources presumably allows both to reduce emissions while continuing to grow their economies, with the assumption that economic growth enhances human well-being. Energy efficiency is also perhaps the only climate policy today embraced by both conservatives and progressives, whether it is cast as buttressing a "green new deal," "energy independence," or "workforce development" strategies (see also a recent debate between Robert Pollin, on one side, and Juliet Schor and Andrew Jorgenson, on the other[11]). And it is not difficult to persuade many parties to accept cost as the ultimate arbiter given that prices have proved to be a powerful mechanism for coordinating economies, especially during an era of ascendant capitalism. Finally, focusing on objective factors like cost and technology seemingly removes politics from decision-making, which is important to bodies like the IPCC that use a consensus-based process to arrive at their conclusions.

While this model-focused approach has proved invaluable in stimulating interest in climate change mitigation, it has recently come under fire. Fellow economists like Nicholas Stern and Martin Weitzman charge that its models fail to capture the uncertainties surrounding the future social costs of climate change and the rate at which to discount such harm.[12] Similarly, the IPCC has begun to express reservations about the adequacy of integrated models, suggesting they are overly simplified, wrongly assume "fully functioning markets and competitive market behaviour," and ignore "social and political forces that can influence the way the world evolves."[13] Other economists defend these models, claiming that while they may omit major risks associated with climate change, they all point to the same conclusion that carbon emissions cause substantial damage and immediate action should be taken to reduce them.[14] They argue that by extending such models to include a wider range of social phenomena, researchers can greatly improve their utility and accuracy. Meanwhile, some industry groups, conservative politicians, and other climate change deniers have pounced on these disputes to suggest all efforts to model the causes or costs of human-caused carbon pollution are useless and, therefore, should be abandoned.

BROADENING THE SOLUTION SPACE AND STIMULATING DIALOGUE

We stand with those who recognize the urgency of anthropogenic climate change. We also support calls to break down disciplinary silos and to conduct more research on the social actors and conditions responsible for and affected by carbon pollution.[15] In that spirit, we note some additional limitations of the conventional modeling approach that we, as environmental sociologists, think warrant special attention.

First and foremost, too often lost in this approach's "big picture" economy-wide and sectoral analyses are power plants themselves, which are treated as an undifferentiated and anonymous mass. This makes it difficult to discern which of these actors are most culpable for the problem of carbon pollution or to answer this basic question: Whodunit? Proponents of the conventional approach divert attention from extreme polluters in one of two ways. Like the spokesperson for the Taichung plant, they contend that if some plants emit more carbon than others, it is probably because they simply provide more electricity, suggesting that these plants' additional emissions are necessary if the wider community is to reap the benefits of electrical power. Or they argue that because hyperpolluters are statistical outliers, they can be dismissed as conceptually inconsequential.

Other studies, however, challenge this reasoning. They find that not only are there huge disparities in the amount of pollutants emitted across sectors, companies, and especially establishments but also actors that emit at higher levels often do so at higher rates as well. Sociologist William Freudenburg reported, for instance, that DuPont's Freeport-McMoRan facility emitted roughly a third of all toxins from the entire U.S. chemicals sector, and, after accounting for differences in output, this facility still had a higher intensity rate than the typical chemical plant.[16] To disregard such extreme cases, he claimed, would be tantamount to ignoring the rare individuals who commit serious crimes or the very rich who own the lion's share of a nation's wealth.

In addition to minimizing the importance of extreme polluters, the conventional approach limits our understanding of the possible causes and mitigation of these plants' and others' emissions. In particular, it erroneously assumes that market and technical solutions address the full range of factors driving carbon emissions, always have their intended effects, are the only viable means to mitigate pollution, and operate best in the absence of citizen activism.

In stressing the primacy of market and technical forces, the conventional approach effectively abstracts power plants from the societal contexts in which they are embedded. Its practitioners may occasionally account for a nation's population size, but other features of societies are largely left unexplored. Hence, this approach precludes an examination of the noneconomic conditions that differentiate plants and might shape their pollution behavior. It is unclear whether net of cost factors, for example, plants owned by dominant utilities or located in countries that figure prominently in the world economy emit more pollutants and, if so, whether these noneconomic factors independently or jointly shape plants' pollution behavior. More simply, context matters.[17]

With respect to its solutions' intended effects, the conventional approach not only promises that improvements in power plants' technical efficiency will reduce their emissions by lowering operational costs but also rejects the possibility that enhanced efficiency could have the opposite effect of increasing emissions by stimulating more energy output.[18] As proof that efficiency gains cannot be reversed by such rebounds, proponents of this approach frequently highlight consumer studies showing that, at most, household behaviors before and after the installation of energy-efficient appliances produce rebounds between 10 and 60 percent but never over 100 percent, the so-called backfire effect.[19] They also justify whatever rebounds occur among low-income households in developing countries as an indication that these countries are achieving development goals and improving productivity. They stop short, however, of directly testing their claim that efficiencies reduce emissions at the plant level where electricity actually is generated and carbon emissions are released.

Finally, in presuming there is a market or technical fix for carbon pollution, the conventional approach not only limits the types of policies that are considered but also fails to determine which *do*—as opposed to *should*—reduce emissions and how mobilized citizens shape those outcomes. Its policy prescriptions are often based on what proponents consider to be objective economic criteria, such as facilitating the expression of consumer preferences. Instead of examining whether the policies recommended actually do better than others at reducing plants' emissions, therefore, supporters of the conventional approach typically advocate those policies regardless, as a matter of principle. By the same token, they discourage outside interventions in the form of citizen pressure, which they claim disrupt free market dynamics.

We seek to address these shortcomings by advancing a comprehensive sociological understanding of electricity-based carbon pollution. This approach redirects attention to what we call society's *super polluters*, foregrounds social factors and their synergistic effects on plants' pollution behavior, examines the conditions under which efficiency measures might actually backfire and cause power plants to emit more carbon pollution, and tests what the real emission effects are of a variety of climate and energy policies and how these effects vary depending on citizen activism. In proposing this alternative approach, our goal is not to replace the reigning social science view drawn from energy and environmental economics. Rather, we hope to broaden its solution space and stimulate dialogue with it as well as other scientific approaches that have largely defined global climate change research. In other words, we seek to foster more cross-disciplinary research by focusing on a material outcome that existing scholarship identifies as urgently important (reducing electricity-based carbon emissions) and suggesting how sociology can further illuminate its wrongdoers, drivers, and reduction.

Importantly, as the climatic dangers of burning fossil fuels become more apparent and scrutiny of utilities that rely on these inputs intensifies, public interest in hyperpolluting facilities has grown. News outlets and organizations like the Weather Channel and the Center for Public Integrity regularly list the U.S. power plants that emit the most carbon pollution.[20] According to the Union of Concerned Scientists and others, these types of report cards speak to how broader societal discussions about responsibility for climate change are shifting the focus from the consumers of carbon-intensive items and services to their producers,[21] in much the same way that cancer caused by cigarette smoking was initially blamed on individuals' lack of self-control but later attributed to cigarette companies themselves. In short, who are the "villains" of climate change?

This development and the fact that carbon emissions from power plants and other sources reached a record high in 2018 and are projected to continue as demand for oil and natural gas rises[22] suggest the time is ripe for a different approach. Toward that end, we expand the current societal interest in U.S. super polluters to include fossil-fuel power plants throughout the world. We also explore the causal relevance of several social factors for plants' carbon releases, examine the actual effects of efficiency measures on plants' overall emissions, and test the efficacy of local abatement policies and citizen mobilization in the absence of federal action.

In doing so, we hope to dispel the perception partly perpetuated by the conventional approach that climate change is due to impersonal market and technical forces over which society has little control. And we point to a potentially more feasible and effective strategy for tackling the world's carbon emissions.

Instead of creating absolute standards for all power plants in a country, we suggest we could get more bang for the buck by targeting plants that produce an egregious portion of a nation's total electricity-based emissions.

THE STRANGE ABSENCE OF SOCIOLOGY AND ITS CONSEQUENCES

How societies are "energized" largely determines their structure, character, status, and ultimate fate.[23] Social life is increasingly understood in terms of phenomena that exhibit energy—becoming rather than being, movement rather than stasis, dynamism rather than fixity. As Ernst F. Schumacher observes, "There is no substitute for energy. The whole edifice of modern society is built upon it . . . it is not 'just another commodity' but the precondition of all commodities, a basic factor equal with air, water, and earth."[24] Even in the so-called information age, information itself is more and more characterized as being dynamic and energetic.

Energy can also be a "great divider." Relations of power within societies depend heavily on which energy forms are dominant and who controls and provides them.[25] In addition to shaping relations within societies, systems of heating, powering, and moving objects can transform power dynamics across them.[26] The emergence of "fossil-fuel societies" in the early eighteenth century, for example, triggered the relative decline of the non-fossil-fuel economies of India and China from that point onward. By the same token, as these and other societies have changed through time, their energy systems have developed various "lock-ins" that prevent human intervention; they combust massive amounts of fossil fuels, which puts them at risk of overshooting their carrying capacities and could lead to momentary "states of barbarism"[27] or total "collapse."[28] Modern, large-scale energy systems, in fact, deployed more energy during the last century than all of human history before 1900.[29] If continued, this trend could produce abrupt energy shortages and catastrophic decline, analogous to those experienced by the Roman Empire and Mayan civilization.

Despite energy's pivotal role in society, sociologists have paid relatively limited attention to the existential threat currently posed by electricity-based carbon pollution and climate change more generally. As an illustration, we searched for "climate change" in three indices in the popular academic search engine EBSCO. Environment Complete, which includes the environmental sciences, returned 138,533 results. EconLit, the American Economic Association's database of more than a million publications, returned 12,337 hits. SocINDEX, which comprises most of sociology, returned only 4,870 results, or less than 4 percent of what is found in the environmental sciences. Why has a discipline devoted to understanding society largely overlooked one of its greatest physical and existential threats?

It may be that sociologists simply find the topic less interesting than others. Substantively different social problems must compete to get on the various public agendas (as represented by, e.g., newspapers, foundations, and congressional committees). And regardless of their seriousness or the number of human lives they threaten, problems that can be more easily dramatized or that align with cultural and political concerns are most likely to win this competition.[30] Applied to sociology, this would suggest that its researchers and audiences have tended to ignore the climate-disruptive emissions of power plants because other important issues are more buzzworthy or resonate more with lifestyle and legal concerns.

Put differently, utilities are just plain boring compared to other important topics that have recently captured the attention of the discipline, such as poverty and affordable housing,[31] the policing of inner-city neighborhoods,[32] and school systems that reproduce privilege.[33] This has occurred despite the fact that climate change, unlike any other problem that society has faced, transforms the relationship between human beings and the basic conditions of life on earth. While the entire range of progressive issues is certainly worthy of investigation, even the greatest victories in the areas of housing, criminal justice, education, and so forth will be ashes in our mouth unless they are won in the larger context of averting climate catastrophe.[34]

Another possibility is that sociologists do not address issues related to climate change because of when they were professionally socialized and how they were taught to theorize "social" problems. Most scholars in the discipline's higher-status departments may have matured before climate change had become a major issue and as a result train their students in the traditional topics of sociology, continuing the disciplinary dominance of those topics.[35] It might also be that consistent with the writings of Emile Durkheim, one of the discipline's founders, sociologists are typically trained to conceive humans as exempt from ecological constraints and thus to confine their studies to social as opposed to physical facts, even though the subfield of environmental sociology has attempted to systematically debunk this anthropocentric approach for decades.[36]

Sociologists may also refrain from studying climate change because of the hostility of policy makers. It would be hard to overstate the degree to which some politicians and legislators have been hostile or unreceptive to social sciences like sociology. This lack of receptivity has discouraged work on a variety of topics, just as the generalized receptivity of policy makers to sociology in the post–World War II era led to something of a golden age of policy-relevant work in sociology on issues including juvenile delinquency, educational inequality, and race relations. Finally, it could be that sociologists do not feel qualified to investigate climate change because very few possess the kind of rudimentary knowledge of climate science that would seem a prerequisite for doing interdisciplinary work on the topic.[37]

Whatever the reasons, the consequences of sociologists not analyzing power plants' carbon emissions seem clear: sociologists are teetering dangerously close to a new form of climate denialism, or perhaps climate avoidance, and could ironically be deemed irrelevant to arguably the single biggest danger to human society. This could occur as social scientific research on climate change continues to be dominated by energy and environmental economists, who tend to study it solely in terms of cost and technical factors. Society could thus be deprived of the sociological imagination needed to address the reality of climate disruption and tackle the energy sector's hyperpolluters.

Fortunately, some sociologists are beginning to address the discipline's neglect of climate change and the environmental impact of energy production. They have revisited founding theorists, such as Karl Marx and Max Weber, and discovered they were quite interested in how capitalism creates a rift between social and natural systems[38] and how social meanings are often "anchored in biophysical realities," including climate change, resource consumption, and energy scarcity.[39] Others have theorized about the energy-society interface,[40] suggesting that the types of technology adopted and their climatic impacts will depend on preexisting social structures, values, geographical conditions, and especially power inequalities.[41]

Building on these contributions, sociologists have generated valuable insights into the effects of different international relationships on emission outcomes.[42] For example, world-systems scholars contend that a nation's carbon discharges are a function of its location in the core, semiperiphery, and periphery zones of the world-system,[43] whereas world society researchers argue that emissions are shaped by a nation's embeddedness in a global pro-environmental order, reflected in its participation in international environmental organizations.[44] Others suggest that pollution is influenced by national political-legal systems that structure the power relations among business, the state, and the citizenry.[45] Still others argue that size, age, and other attributes of organizations shape their propensity to pollute[46] and document the strategies used by the powerful fossil-fuel industry to contest scientific evidence about climate change.[47] Treadmill of production and ecological modernization scholars have likewise debated whether improving the efficiency of energy technologies is enough to compensate in the long run for the harmful ecological effects of scale.[48] Finally, sociologists have questioned the adequacy and appropriateness of the Kyoto Protocol and other top-down approaches to climate change, arguing that environmental policy making must be attuned to local differences and design contextually sensitive measures.[49]

While these inquiries suggest that sociology is slowly becoming sensitized to the threat posed by climate change and to the need to decarbonize electricity, they nonetheless stop short of examining which power plants emit the most carbon and why. When describing environmental impacts, these studies, like those of energy and environmental economists, portray sectors like electricity in broad strokes and focus on aggregate pollution outcomes, glossing over disparities in plants' emissions. Sociologists have treated world-system location, world society embeddedness, political-legal systems, and organizational factors as competing predictors of pollution rather than exploring how they might work in concert to shape power plants' emissions. Treadmill of production and ecological modernization researchers theorize that environmental degradation is shaped by upstream producers like utilities, but their empirical analyses of the pollution effects of efficiency focus on the downstream consumers who purchase the electricity produced. And despite their preference for locally crafted climate and energy policies, sociologists have not examined whether subnational measures actually curb plants' emissions and, if so, under which communal conditions they are most effective. In short, sociology has yet to take the next step and develop a systematic alternative framework for identifying egregiously destructive plants and explaining the causes and mitigation of their carbon emissions.

BRINGING SOCIOLOGY BACK IN: A SUPER POLLUTERS APPROACH

In this section, we take that next step and sketch the contours of what we label a *super polluters* approach. To explain how this framework differs from energy and environmental economists' conventional approach, we contrast what each says and predicts about four issues that comprise this book's substantive chapters: disproportionalities in the sources of pollution, the causal relevance of social conditions, the intended effects of efficiency, and the efficacy of local policies and activism. And we briefly suggest how the analytical strategies that these approaches apply to the problem of anthropogenic carbon pollution diverge.

Disproportionalities

Unlike mainstream economists, who have long been interested in variations in organizations' externalities and productivity,[50] energy and environmental economists tend to whitewash such differences in summing emissions to the sector level. Consistent with a rational actor paradigm, their conventional approach assumes all economic actors, persons or organizations, are calculating, self-interested, instrumental maximizers with fixed preferences. It posits that a single motive (e.g., providing a reliable and affordable service) can usually explain most of the environmental behavior and output of certain sets of actors (e.g., power plants). By aggregating these similarly motivated actors, proponents of this approach can build up to higher levels of analysis (e.g., sectors, entire economies) and compare summed pollution outcomes. Although there may exist variation in pollution levels across sources, they disregard these because the same motive that drives pollution presumably also drives productivity, causing the former to increase in rough proportion to the latter. Thus, the conventional approach diffuses responsibility for externalities and precludes the possibility of implicating any single actor or set of actors for most of the environmental harms associated with a sector.[51]

While environmental sociologists have traditionally stressed the constraints and opportunities that social systems operating at the macro level create for actors and the national differences in environmental behavior tied to uneven development and international inequality,[52] they have increasingly cast their analytical focus downward to the meso level and studied subnational disparities in pollution outcomes. Initially, they examined differences in experiencing environmental harms, documenting how poor and minority neighborhoods are susceptible to industrial pollution.[53] More recently, they have investigated polluters themselves and differences in the production of environmental harms,[54] finding that some actors are profligate polluters, while others are not.[55] And rather than dismissing these disparities as artifacts of proportional increases in output, they contend major polluters often have privileged access to natural resources like fossil fuels that allows them to produce commodities increasingly at the expense of the environment.[56] In keeping with this logic, a super polluters approach decomposes sectors to illuminate differences in the production of harm and expects some actors not only to externalize more pollutants but also to do so at a level disproportionate to their productivity.

Simply stated, whereas the conventional approach suggests a few actors will pollute at extreme levels because they generate commensurate amounts of output, *a super polluters approach predicts the same actors that pollute at high levels will also tend to pollute at high rates.*

Social Conditions

Proponents of the conventional approach seek to develop parsimonious models that predict, demonstrate, and explain "rational behavior." With respect to climate change, this usually means restricting predictors to a few economic factors. Variations in social conditions are commonly treated as irrational "noise" or unsystematic biases that can be eliminated by theoretical generalizations or filtered out via error terms or fixed effects in regression models. In this way, actors are basically treated as socially disconnected decision makers.

The rational actor perspective that informs the conventional approach has been criticized for assuming the preeminence of a utilitarian logic and ignoring the influence of other nonrational factors on behavior. The proponents of this perspective do not deny that other preferences may exist and motivate actors to maximize their self-interests, but they rarely examine them, suggesting such preferences lie beyond the scope of their analysis. Economics as a whole, however, has long since moved beyond the rational actor approach or at least tempered its most extreme assumptions in acknowledging a wider range of behavioral assumptions and developing more sophisticated models of human and organizational behavior.[57] Applied to the problem of climate change, some of these newer frameworks, like agent-based modeling, can suggest, for example, how differently motivated actors interact to generate environmental norms and forecast the emission scenarios of particular societies.[58] Still, like the conventional approach, these frameworks struggle with problems of behavioral realism.[59] It is unclear, for instance, how we should test their simulations' robustness using real-world data.[60] They also tend to rely on crude proxies like gross domestic product and population size to represent complex societal forces and structures.[61]

Sociologists have long argued that economic actors do not operate in a social vacuum. Rather, they are embedded in societal complexes and are thus subject to a variety of noneconomic pressures.[62] In this vein, as noted earlier, environmental sociologists have explored the effects of various exchange, cultural, political, and organizational factors on emission outcomes. In addition to suggesting how these structures shape pollution behavior above and beyond the effects of prices, these scholars stress how they are often elements of larger social complexes, further entrenching the drivers of pollution. We argue the same is likely true with respect to electricity-based carbon emissions.

That is, whereas the conventional approach assumes social conditions are too random to have a significant effect on emissions net of price factors, *a super polluters approach predicts such conditions will have not only significant independent effects but also synergistic ones.*

Efficiency

According to the conventional approach, organizations seek to optimize their assets by investing in technologies that use fewer resources to provide the same service. Power plants, for instance, will seek to minimize the amount of fuels they use to generate electricity for their customers. By reducing the costs of production, energy efficiency technologies enable organizations[63] to simultaneously expand their operations and reduce the pollution associated with their inputs.[64]

Those using the conventional approach put considerable effort into trying to identify obstacles to increased efficiency. Typically, they attribute unwillingness to exploit apparently profitable energy efficiency opportunities to a variety of "hidden costs," or market failures such as the landlord-tenant barrier. But they stop short of addressing more serious "distortions" or the possibility that efficiency gains might rebound to some extent and produce unintended negative consequences, such as backfire effects. This oversight is ironic given that economists like William Stanley Jevons[65] were among the pioneers of rebound research who attributed such phenomena to common economic dynamics. It is also ironic considering that the same natural systems approaches with which they seek alignment stress how human and natural systems are subject to complex feedback loops that can create harmful and irreversible outcomes.[66] These discrepancies between what theory says and what energy and environmental economists commonly practice suggests that either the conventional approach overstates the price elasticity of energy usage or its analysis is obscured by a normative agenda.

Since Max Weber first wrote on the potential contradictions of rationalization,[67] sociologists have been skeptical of the promised benefits of efficiency. They note, for example, that efficiency measures designed to reduce energy service costs often free up resources that consumers, in turn, increasingly purchase, causing emissions associated with those resources to also rise.[68] From a super polluters perspective, similar backfire effects may occur upstream when the efficiency measures taken by actors like power plants are not sufficient to compensate for the harmful ecological effects of subsequent growth in electrical output. This is especially likely when plants are older, larger, and subject to social pressures that make changes to existing environmental practices difficult or risky.[69]

Whereas the conventional approach argues that efficiency measures will improve carbon emissions, *a super polluter approach predicts that they will have the opposite effect of increasing pollution, especially among actors prone to inertia.*

Local Policies and Activism

The policy prescriptions of the conventional approach tend to be highly normative in the sense that they prioritize market-based strategies. Those who support these prescriptions recognize that regulations are sometimes necessary but prefer creating market incentives to achieve them because the latter are presumably more efficient. Their goal, therefore, is to perfect policy options by showing how such incentives can be aligned in ways that strongly encourage compliant behavior. They assume that policies incentivized and designed according to market principles will, in fact, be effective after they are implemented, provided investment environments are not disrupted by civil unrest or pressure.

However, as institutionalists in sociology observe,[70] even the best designed rules and policies may have little effect if they are decoupled from actual practices. Organizations, including government agencies, frequently refrain from enforcing formal requirements and engage instead in ceremonial gestures to legitimate themselves in the eyes of the public. Only when the public holds organizations accountable for their promises is real change likely to occur.[71] Economic sociologists have been quick to incorporate these insights,[72] but they are still largely overlooked by energy and environmental economists.

Consistent with institutionalists' arguments, a super polluters approach is ambivalent about the free market. Its proponents take a realistic approach to policy, leaving open the possibility that any existing measure may or may not succeed in reducing emissions. And they treat policies less like incentives than fragile compromises that require constant social pressure to be effective and preserved.[73]

It follows that whereas the conventional approach argues that only market-based policies like efficiency measures can decrease emissions and that market interventions in the form of citizen pressure undermine these policies' effectiveness, *a super polluters approach predicts that nonmarket policies can be effective as well and that the success of efficiency programs is actually enhanced by citizen engagement.*

Analytical Strategies

In general, energy and environmental economists take a "clean models" approach to the problem of anthropogenic carbon pollution,[74] which offers predictions about the causes and mitigation of emissions deduced from a small set of assumptions about the motivations of actors. This conventional approach is also clean in the sense that its models generate forecasts or scenarios that are rarely tested and validated with historical data.[75] Sociologists, in contrast, subscribe to a "dirty hands" approach, which is more empirical (as opposed to formal modeling), backcasts or scrutinizes the veracity of hypotheses using historical data, and refrains from imposing what sociologists consider a relatively superficial elegance on messy problems.

LOOKING FORWARD

Until recently, it was difficult to apply this sociological approach for super polluters to the problem of facility-level carbon pollution due to a lack of comparable data on plant-level emissions. However, this obstacle has been removed with the recent development of data sets like the U.S. Environmental Protection Agency's Greenhouse Gas Reporting Program (GHGRP) and the Center for Global Development's Carbon Monitoring for Action (CARMA) initiative, which, respectively, allow researchers to analyze and compare for the first time the determinants of CO_2 emitted by individual power plants in the United States and in other nations throughout the world. As other data sources like these, such as information gathered through satellite monitoring systems, become more available, it will be increasingly incumbent on researchers to take a dirty hands approach to the study of electricity-based carbon pollution.

In the next chapters, we use the GHGRP and CARMA files to address a series of sociologically grounded questions about this problem that until now have largely gone unanswered. These questions by no means exhaust all that could be raised by our super polluters framework. Sectors other than electricity that emit disproportionate amounts of CO_2, such as agriculture/land use and transportation, could also be the subjects of inquiry.

We address a policy that enjoys the support of both the left and the right (energy efficiency) as well as policies that have been widely adopted at the subnational level (e.g., renewable portfolio standards), but there are more divisive ones (e.g., nuclear energy), and others are on the horizon (e.g., geoengineering, carbon capture and storage) that are equally deserving of attention. And there are additional "villains" that could be investigated besides the hyperemitting power plants we discuss, ranging from the wealthy households that create demand for fossil-fuel goods[76] to the manufacturers that transfer vast quantities of emissions via international trade[77] to the major suppliers and financiers of fossil fuels.[78] Hopefully, the questions we raise here and the tentative answers we provide to them in subsequent chapters are suggestive of the new empirical directions that research on the sources, causes, and mitigation of climate change could take if it brought society and its super polluters to the fore.

Chapter 2 examines the extent to which carbon pollution associated with electric utilities can be explained by the behavior of a small set of facilities. Here we ask, To what extent are plant-level carbon emissions distributed unevenly within countries' electricity sectors? Is there a relationship between such disproportionalities in plant-level emissions and higher overall electricity-based emissions? Could countries substantially lower their electricity sectors' carbon emissions by targeting extreme polluters located within their borders?

Chapter 3 illuminates the social sources of electricity's carbon emissions. We use a theoretically integrative and multimethod framework to address these questions about the structural makeup of the world's hyperpolluting power plants: How might the global, political, and organizational determinants of pollution identified by sociologists combine to cause some fossil-fueled power plants to emit carbon at vastly higher rates than others (net of traditional predictors like fuel type and electricity costs)? Do the same configurations of social structures that increase polluters' emission rates also increase their emission levels? And if so, in which countries are these polluter profiles most prevalent?

Chapter 4 scrutinizes the claim espoused by both the Obama and the Trump administrations that efficiency measures are a viable reduction strategy. Specifically, we seek answers to the following: Do technical improvements in U.S. plants' heat-rate efficiencies reduce their carbon emission rates and levels, or do they inadvertently increase these, as skeptics suggest, by enhancing overall electrical output? What impact does efficiency have on the emissions of plants in other nations throughout the world as well as in the United States, and to what degree does the effect of efficiency on their emissions depend on such noneconomic factors as a plant's size and age and its country's position in the world economy and openness to global environmental norms?

Chapter 5 empirically assesses possible policy solutions by posing and answering the following questions: How effective are subnational climate and energy policies, like those developed in the United States, at reducing power plants' emissions? How do market-based policies such as cap-and-trade compare with others like electric decoupling that guarantee utilities revenue regardless of their productivity? Is the effectiveness of policies enhanced or undermined by local movement and civil society infrastructures in the form of environmental nongovernmental organizations? If enhanced, what are the mechanisms through which these infrastructures reduce power plants' carbon emission levels? Because this chapter includes quantitative analyses of the effects of policies and activism and qualitative case studies of local activist groups that have shaped the effectiveness of policies, it is somewhat longer than the preceding chapters, which perform quantitative analyses only.

Chapter 6, our final chapter, takes stock of our findings and their relevance for our super polluters framework and discusses their shared implications for the development of a future research and policy agenda.

CONCLUSION

The question of who is to blame for anthropogenic climate change has animated public and policy discourse since the United Nations Framework Convention on Climate Change established the principle of "common but differentiated responsibilities." According to the latter, nations that produced the most carbon emissions in the past are most responsible for avoiding dangerous anthropogenic interference with the climate in the future. But as nations faltered in meeting their commitments to global net-zero GHG gas emissions toward the end of the twentieth century, discussions about climate responsibilities have slowly shifted and focused on nonstate actors. Who is ultimately held responsible for carbon pollution will be for society to decide. Our study and framework seek to inform that decision by shining a new, analytical light on those nonstate actors that currently spew the most carbon pollution—super polluters—in the face of scientific evidence that such emissions imperil our planet's atmosphere.

In addition, our work advances a burgeoning literature on "problem-solving sociology."[79] Whereas past attempts to demonstrate the relevance of the discipline to contemporary social problems have described and decried social ills, studied the experiences of their victims, and/or critiqued existing fixes,[80] this new strand of scholarship strives to identify root causes, investigate culprits, and propose new solutions. In the same vein, we offer a novel sociological perspective on the determinants, villains, and abatement of perhaps the most wicked[81] problem of all—society-induced climate change. We view this as a complement to other important research documenting the actions of groups and organizations that resist climate mitigation.[82]

In proposing this perspective, we are not suggesting that cleaning up the dirtiest power plants will solve the problem of climate change. We contend that targeting these lowest-hanging fruits is a necessity given that societies need to radically transform their energy systems in little more than a decade to keep global temperatures from rising more than 1.5°C and thus avert severe climate impacts.[83] At the same time, we recognize that this strategy alone is by no means sufficient and that society must consider several options capable of delivering high impacts in a short amount of time, including the full adoption of proven green technologies[84] and programs initiated by for-profit businesses to change their own and their trading partners' environmental behavior.[85] Nor do we pretend that our study is the final word on hyperemitting power plants and other types of super polluters. Rather, our hope is to open up discussion about climate culpability and spur further research on the chief perpetrators of climate change that may even challenge our findings and framing of the issue. Importantly, in adopting the term *villain* to describe these polluters, our intent is not to demonize particular actors but to address the diffusion of responsibility surrounding the problem of climate change. Finally, our aim here is not to provide a comprehensive review and integration of the vast economic and engineering literatures on power plants and other carbon polluters. We seek instead to show how the underutilized discipline of sociology can help bring society back into the study of anthropogenic climate change and thus facilitate a more truly interdisciplinary approach.

2

CLEANING UP THEIR ACT

Potential Emission Reductions from Targeting the Worst of the Worst Power Plants

A TALE OF TWO PLANTS

Perhaps no national electricity sector has both inspired and puzzled clean energy advocates more than that of Germany. A 2015 story in *National Geographic* described the country as "pioneering an epochal transformation it calls the *energiewende*—an energy revolution that scientists say all nations must one day complete if a climate disaster is to be averted."[1] Then, in the early morning of January 1, 2018, for the first time ever, nearly 100 percent of Germany's energy was supplied by renewable sources, a historic achievement for one of the world's largest economies. Nearly 85 percent was supplied by wind alone, assisted by unusually strong early morning gusts and, according to some reports, a country still sleeping off its New Year's celebrations. Over the course of 2018, a combination of onshore and offshore winds supplied about 15 percent of Germany's power mix, and all renewables generated more than one-third.[2]

Still, the day turned out to be largely symbolic. Germany has since delayed its installation of wind turbines, and it is not clear whether it will be able to meet its own emissions targets. In 2007, German chancellor Angela Merkel pledged that Germany would reduce emissions by 40 percent by 2020 compared to 1990 levels. Recent estimates, however, reveal that the country has reduced emissions by only slightly more than 27 percent and, according to a leaked report from its Ministry of Environment, is on track to reduce carbon emissions by 31.7 to 32.5 percent by 2050—far short of the goal of 80 to 95 percent set as part of the Paris Agreement to address climate change.[3]

One reason is Germany's reliance on coal[4]—and dirty coal, at that. For hundreds of years, the country has mined and burned lignite, the so-called brown coal that emits more carbon dioxide (CO_2) than hard coal and nearly twice as much as natural gas. And as renewables have supplied more energy for the sector and dropped overall energy prices, lignite has simply been more affordable to burn. Take, for example, the Niederaussem lignite-fired plant near the western city of Cologne, which, in 2009, emitted 26.3 million tons of CO_2 and produced nearly 28 billion kilowatt-hours (kWh) of energy. The plant was listed in the World Wide Fund for Nature's 2013 *Dirty Thirty* report on the most polluting power plants in Europe. In our data, the Niederaussem plant ranks as the seventh-dirtiest plant in the world.

Just behind Niederaussem is the Jänschwalde lignite-fired plant, on the German-Polish border, which was not quite as dirty but was more intense, emitting 26.3 million tons of CO_2 while generating slightly more than 21 billion kWh of energy. It was the world's sixteenth-dirtiest plant in 2009. At the other end of the spectrum in Germany is the Brunckviertel natural gas plant in Ludwigshafen am Rhein. Of the more than one thousand fossil-fuel-burning power plants in Germany that year, the Brunckviertel plant was the cleanest, emitting 4.58 tons of CO_2 but also generating only 6,642 kWh of power. Comparing the Niederaussem and Brunckviertel plants reveals a dispersion of more than 26 million tons of CO_2 between the dirtiest and the cleanest fossil-fuel-burning power plants within Germany. In fact, the Niederaussem and Jänschwalde plants alone accounted for more than 16 percent of the total carbon emissions of the German energy sector, and the country's ten dirtiest plants (less than 1 percent of all power plants) accounted for nearly 45 percent.

Figure 2.1 maps out the locations of the top ten polluting power plants in the world in 2009. We see that the first and the eighth are located in Taiwan; the second, third, fifth, and sixth are in South Korea; the fourth is in Poland; the seventh, as already mentioned, is in Germany; the ninth is in India; and the tenth is in South Africa. To understand the climate impact of the energy sector in Germany or any other country, it might be instructive to focus on a small subset of dirty power plants within it.

ELECTRICITY AND THE ISSUE OF DISPROPORTIONALITY

The combustion of fossil fuels for the production of electricity comprises the single largest contributor to sector-level, human-caused greenhouse gases (GHGs), accounting for roughly a quarter of all emissions.[5] Over the past two decades, the electricity sector's CO_2 emissions have risen by 60 percent worldwide.[6] The Intergovernmental Panel on Climate Change recently suggested that CO_2 emissions from this sector could double from 2010 baseline levels by 2050[7] given the growth in the number of power plants throughout the world, especially in rapidly developing nations where fossil-fuel-burning power plants make up a large portion of the electricity generation sector.[8]

Given these challenges, many researchers and policy experts have concluded that focusing on reducing the emissions of

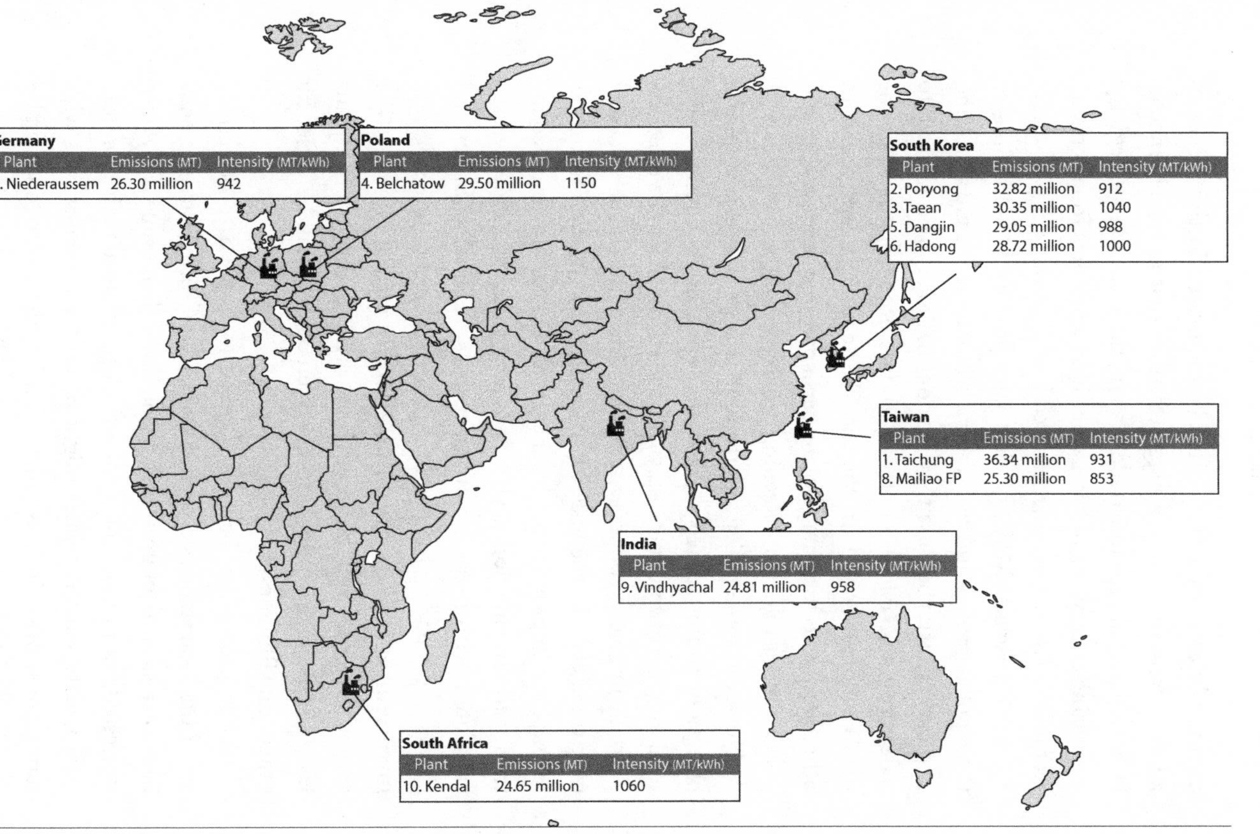

FIGURE 2.1 Top ten polluting power plants in the world, 2009.

countries' electricity sectors could be the most effective means to a low-carbon future. They suggest further that in the absence of an international cap-and-trade system, the next best strategy for decreasing energy-related emissions is to set targets for electricity sectors' carbon intensities, measured by the amount of CO_2 emitted per megawatt (MW) of electricity produced.[9] According to these policy groups and researchers, sectoral targets are appealing to most countries because they are nonbinding, can be adjusted to reflect best available technologies, and do not set limits on countries' economic growth in the way that emission caps could. In other words, sectoral targets are perhaps more manageable and politically feasible than sweeping economy-wide approaches.

However, some policy and research groups have raised several objections to this strategy.[10] First, they contend that because sectoral targets are voluntary, they lack the stringency to be effective. Second, in many countries, emission intensities have declined over time, even in the absence of targets, due to such factors as technology improvements in energy efficiency. However, countries' carbon emission levels and intensities are only weakly correlated, with many countries, such as the United States and China, experiencing long-term increases in overall levels despite improvement in their intensities.[11] Finally, setting targets at the sector level may not be enough to drive change at the plant level—some plants' efforts to lower emissions may be offset by others' unwillingness to do the same.[12]

While proponents have been quick to respond to objections about the legal status, stringency, and form of sectoral targets, they have been slow to address the last one about their scope. They have suggested, for instance, how countries can improve their monitoring capacity, what baselines might be used for comparisons with future emissions, and how targets may be adjusted to lower emission levels.[13] One example is the 2015 Clean Power Plan in the

United States, which gave each state options for cutting its emissions through a combination of energy efficiency improvements and investments in renewables and natural gas and nuclear power generation. However, they have yet to explore differences in the levels and rates at which individual fossil-fueled power plants emit carbon. This is despite William Freudenburg's sociological research on disproportionality in the production of pollution among manufacturing facilities within the United States.[14]

Freudenburg hypothesized that a small subset of a sector's facilities is often responsible for the lion's share of its toxic emissions. Using data from the Toxic Release Inventory, a publicly available database containing information on facility-level toxic chemical releases in the United States, he calculated Gini coefficients to quantify disproportionality in facility-level pollution.[15] The Gini coefficient is a widely used measure of inequality and can range in value from 0 (perfect equality, or perfect proportionate distribution) to 1 (perfect inequality, or perfect disproportionate distribution).

Freudenburg found that there is substantial inequality in emissions and that such disproportionalities in toxic releases are modestly amplified when industries are normalized by the number of employees and payroll expenditures, suggesting that varying levels of pollution are not simply the result of higher levels of production in some industries versus others.[16] He concluded that toxic pollution within the United States could be mitigated significantly if a relatively small fraction of producers substantially reduced their pollution and that such environmental improvements would not greatly disrupt the overall economy or threaten industry survival.

In a recent study, Anya Robertson and Mary Collins[17] adopted Freudenburg's approach and conducted a disproportionality analysis of the generation of CO_2 emissions in the U.S. coal-fired electric utility industry. Through the calculation and examination of Gini coefficients, they found that facility-based disproportionality patterns are largely attributable to the amount of electricity a power plant generates. Their study of U.S. coal plants speaks to the value of applying the disproportionality framework and analyzing the energy sector's carbon emissions using Gini measures. However, the study does not consider all fossil-fuel plants within the United States and elsewhere.[18] It is unclear, therefore, whether disparities in plants' emissions are simply a function of their electrical output or depend on the types of fossil fuels they use and the countries in which they reside.

In this chapter, we extend sociological research on disproportionality by focusing on potential disproportionality in carbon emissions for fossil-fuel power plants in nations across the world. However, first we provide a descriptive analysis of the distributions of plant-level emission rates and intensities for the ten nations that have the highest national levels of emissions from the electricity sector. The emissions data used have been obtained from the Center for Global Development's Carbon Monitoring for Action (CARMA) database.[19] Such an exploration will help determine if a disproportionality analysis of the world's power plants could be useful. These data are presented for 2009, the most recent year for which they are currently available for all countries.

A DESCRIPTIVE ANALYSIS OF TEN NATIONS

Columns 1 and 2 of table 2.1 report the overall emission levels (millions of tons) and rates (millions of tons of carbon per unit of electricity produced) of the ten countries with the most CO_2 emissions from the electricity sector in 2009. Column 3 shows what share of a country's electricity-related carbon emissions

TABLE 2.1 TOP TEN CO_2-EMITTING COUNTRIES, 2009: ALL POWER PLANTS

Nation	Electricity Sector's CO_2 Emission Level (millions of tons)	Electricity Sector's CO_2 Emission Rate	Top 5% Share of Sector's Total CO_2 Emissions (%)	Top 5% Share of Sector's Total Electricity Generated (%)	Emission Rate Ratio of Top 5% to Average Plant	% Electricity from Fossil Fuels	Primary Fossil Fuel
China	2,840	823	37	31	1.92	80	Coal
United States	2,320	609	75	47	1.42	69	Coal
India	694	835	75	58	2.85	81	Coal
Russia	542	587	45	29	1.41	65	Gas
Japan	387	405	97	63	1.83	60	Gas
Germany	303	569	98	64	5.11	63	Coal
South Korea	233	549	84	52	1.63	65	Coal
Australia	219	907	93	83	2.07	93	Coal
United Kingdom	175	497	96	75	1.92	66	Gas
Saudi Arabia	161	800	53	56	0.83	99	Liquid

can be attributed to the top 5 percent of polluting power plants. In China's electricity sector, the top 5 percent account for over a third of total carbon discharges. In the next two most heavily polluting countries—the United States and India—the top 5 percent are responsible for nearly double that amount (or three-fourths of emissions). And in the remaining countries, the top 5 percent account for between 45 percent and 98 percent of national carbon emissions from the electricity generation sector.

While the uneven distribution of carbon emissions partially reflects the fact that the top 5 percent also generate a large share of their sector's total electricity, as the numbers in column 4 indicate, it is still the case that, except in Saudi Arabia, these plants emit more carbon per unit of electricity produced than the average plant, as shown in column 5. In China, for instance, the worst polluters emit carbon at a rate 1.92 times greater than that of the average plant.

The type of fuel used by these plants is likely an important determinant of their higher emission rates, as is the extent to which they burn fossil fuels. Column 6 indicates that for each of these ten nations, the percentage of electricity generated from the burning of fossil fuels ranges from 60 percent in Japan to 99 percent in Saudi Arabia. And column 7 indicates that the primary fossil fuels used by power plants vary to some extent across these nations, with coal being the primary fossil fuel for plants in China, the United States, India, Germany, South Korea, and Australia. Gas is the primary fossil fuel for the plants in Russia, Japan, and the United Kingdom, while for Saudi Arabia, it is liquid fossil fuels.

Table 2.2 narrows the scope to only fossil-fuel power plants in each of the same ten nations. Column 1 reports the emission rate for the fossil-fuel power plants, column 2 reports the top 5 percent's share of the sector's electricity generated from fossil fuels, and column 3 reports the emission rate ratio of the top 5 percent to

TABLE 2.2 TOP TEN CO_2-EMITTING COUNTRIES, 2009: FOSSIL-FUEL POWER PLANTS

Nation	Electricity Sector's CO_2 Emission Rate of Fossil-Fuel Plants	Top 5% Share of Sector's Electricity Generated from Fossil Fuel (%)	Emission Rate Ratio of Top 5% to Average Fossil-Fuel Plant
China	1,018	51	1.02
United States	734	64	1.35
India	1,000	53	1.11
Russia	883	50	1.07
Japan	631	96	1.15
Germany	694	97	1.1
South Korea	766	90	1.02
Australia	770	93	1.17
United Kingdom	593	86	1.04
Saudi Arabia	971	60	0.83

the average fossil-fuel power plant. We see that China's fossil-fuel plants have the highest average emission rate, closely followed by plants in India. The top 5 percent's share of the sector's electricity generated from fossil fuels ranges from 51 percent in China to 97 percent in Germany, with three of the other ten nations having shares or 90 percent or more. Only one of the ten nations has an emission rate ratio of the top 5 percent to the average fossil-fuel power plant below a value of 1 (column 3): Saudi Arabia. For the nine nations having ratios greater than 1, the United States has the largest at 1.35, followed by Australia (1.17), Japan (1.15), and India (1.11). Table A2.1 in the appendix to chapter 2 (at the end of the book) lists the top five polluting plants in each of these ten nations as well as their primary fuel source, their parent company, and their overall carbon emissions in 2009.

Table 2.3 reports the percentage of a country's total electricity-based carbon emissions that would be erased if their top 5 percent of polluting power plants continued to generate the same amount of electricity but lowered their intensities to the average for all power plants in that country's electricity sector (column 1), to the average for all power plants in the world (column 2), to the average for fossil-fuel power plants in that country's electricity sector (column 3), and to the average for all fossil-fuel power plants in the world (column 4).

As column 1 shows, if the top 5 percent of polluters lowered their intensities to the average for their sector, the result would be a reduction of carbon emissions in every country except Saudi Arabia, with these reductions ranging from 13 percent in Russia to 79 percent in Germany. Column 2 indicates that if the same plants lowered their intensities to the world average for power plants, each of these ten countries would reduce emissions by at least 28 percent. Some might object that using such averages is inappropriate because they include power plants using nonfossil fuels. A more appropriate comparison, therefore, might be made using averages for just fossil-fuel plants. Columns 3 and 4 provide these comparisons. In column 3, eight of the ten nations would still see overall emission reductions if the top 5 percent of polluters lowered their intensity to their sector's average for fossil-fuel plants. Virtually the same holds in column 4, where eight of the ten nations would still experience reductions in carbon emissions if the hyperpolluters of the top 5 percent lowered their intensity to the world's average for fossil-fuel plants.

In addition to pointing to a "behavioral wedge" that countries might leverage to hasten emission reductions,[20] these patterns for the power plants in the ten highest-emitting nations support Freudenberg's argument that researchers should study potential disproportionality in facility-level pollution. But a more systematic analysis is needed to assess the degree of disproportionality

TABLE 2.3 PERCENTAGE CHANGES IN ELECTRICITY-BASED CO_2 EMISSIONS IF DIFFERENT CARBON-INTENSITY TARGETS ARE APPLIED TO THE TOP 5 PERCENT OF POLLUTERS IN THE TOP TEN CO_2-EMITTING COUNTRIES

Nation	If Top 5% Lowered Their Intensity to Their Sector's Average for All Power Plants	If Top 5% Lowered their Intensity to the World's Average for All Power Plants	If Top 5% Lowered Their Intensity to Their Sector's Average for Fossil-Fuel Power Plants	If Top 5% Lowered Their Intensity to the World's Average for Fossil-Fuel Power Plants
China	−18	−28	1	-9
United States	−22	−57	−13	−18
India	−48	−58	−8	−24
Russia	−13	−33	−4	-9
Japan	−44	−65	−14	3
Germany	−79	−67	−9	-1
South Korea	−33	−60	−8	-8
Australia	−48	−70	−22	−23
United Kingdom	−46	−58	−4	22
Saudi Arabia	11	−37	11	-3

in the emissions of the world's fossil-fuel power plants and the extent to which such disproportionalities have an impact on overall carbon emissions. These important questions are the focus of the remainder of this chapter.

Next, we examine disproportionalities in the emissions of power plants for 161 of the world's nations. Using facility-level data for fossil-fuel plants across the globe, we calculate national-level Gini coefficients for power plant carbon emissions in 2009.[21] In addition to measuring such disproportionalities in power plant carbon emissions for the majority of the world's nations, we use multiple regression to evaluate whether disproportionalities in plant-level emissions are associated with these nations' overall carbon emissions from fossil-fuel-based electricity production. If we find that national-level emissions from fossil-fuel power plants are associated with disproportionality in plant-level emissions, especially after accounting for the effects of established human drivers of GHG emissions, then targeting extreme polluters in the electricity generation sector could, in fact, be a viable climate change mitigation strategy for many nations throughout the world.

CALCULATING AND EXAMINING DISPROPORTIONALITY GINI COEFFICIENTS

To calculate the disproportionality Gini coefficients for the 161 nations, we used the statistical analysis program Stata[22] and the facility-level measures of the total pounds of CO_2 emitted in 2009 by each of the 19,941 fossil-fuel power plants located in these nations.[23] We weighted each plant by plant-level output, measured as net megawatt-hours generated in 2009, thereby accounting for how efficiently electricity is generated by each fossil-fuel power plant when calculating the Gini coefficient. If we had not used this weighting technique, the disproportionality coefficients would have simply been overly influenced by larger plants that burn more fuel and emit more carbon regardless of their level of efficiency.[24]

Figure 2.2 is a histogram for the estimated disproportionality Gini coefficients for the 161 nations in the study. While a Gini coefficient typically ranges from 0 to 1, we multiplied the coefficient by 100 to aid in interpretation, a common approach when studying such measures. Thus, a value of 100 would indicate that a single power plant produced all of the carbon emissions in a given sector (perfect inequality), whereas a value of 0 would indicate that all power plants emitted an equal amount (perfect equality).

As the histogram indicates, the distribution of the Gini coefficients is very close to normal, which we confirmed statistically

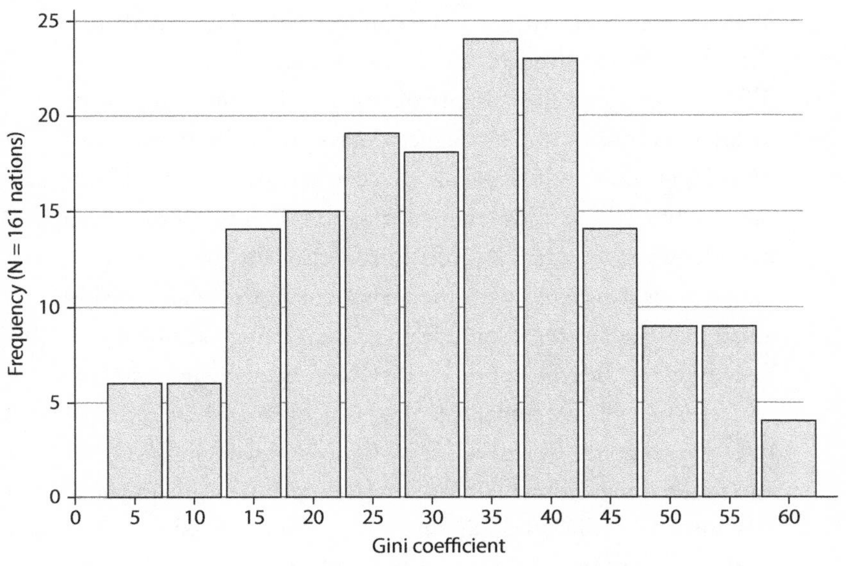

FIGURE 2.2 Gini coefficients for disproportionality in plant-level carbon emissions for 161 nations.

(skewness value = -0.03). The values of the Gini coefficients range from a low of 2.85 (Namibia) to a high of 58.17 (Germany). They have an average of 30.38, a median (50th percentile) of 31.17, and a standard deviation of 12.58, meaning that close to 68 percent of the nations' disproportionality Gini coefficients in our data set range in value from 17.80 to 42.96, while approximately 95 percent of them range in value from 5.22 to 55.54. These descriptive statistics indicate a notable amount of variation in disproportionality in plant-level carbon emissions across nations as well as a relatively symmetrical distribution. More importantly, all nations in the sample are characterized by disproportionality to some extent in their fossil-fuel power plants' carbon emissions.

Table 2.4 lists the disproportionality Gini coefficients for each of the ten nations with the largest national carbon emissions from such power plants for 2009. These are the same nations included in tables 2.1–2.3. Each has a disproportionality Gini coefficient that is above the average for the entire sample of 161 nations. Table 2.4 also lists the number of fossil-fuel plants within each of these nations as well as the percentages of the nations' fossil-fuel plants whose primary fuels are coal, gas fossil fuels, and liquid fossil fuels. This additional information is provided so we can consider whether national-level disproportionality in power plant emissions is associated with the actual number of plants within a nation as well as the composition of fuels being used by fossil-fueled plants. The disproportionality Gini coefficients, numbers of fossil-fuel plants, and percentages of plants whose primary fuels are coal, gas fossil fuels, and liquid fossil fuels for all 161 nations are provided in table A2.2 in the appendix to chapter 2.

China, a rapidly developing nation with the highest national-level emissions from fossil-fuel power plants, has a Gini coefficient of 35.87 for 1,130 fossil-fuel plants, 81.50 percent of which burn coal as a primary fuel. The United States, the second-biggest

TABLE 2.4 GINI COEFFICIENTS FOR DISPROPORTIONALITY IN PLANT-LEVEL CO_2 EMISSIONS FOR THE TEN NATIONS WITH THE HIGHEST OVERALL CO_2 EMISSIONS FROM FOSSIL-FUEL POWER PLANTS, 2009

Nation	Disproportionality Gini Coefficient	Number of Fossil-Fuel Power Plants	% Coal Fossil-Fuel Power Plants	% Gas Fossil-Fuel Power Plants	% Liquid Fossil-Fuel Power Plants
China	35.87	1,130	81.50	8.68	9.82
United States	48.86	2,612	21.75	53.52	24.73
India	46.97	737	37.72	18.86	43.42
Russia	49.46	529	19.28	65.97	14.75
Japan	42.23	1,908	4.61	38.63	56.76
Germany	58.17	965	12.23	71.71	16.06
South Korea	39.61	198	15.15	33.33	51.52
Australia	41.53	434	9.45	47.24	43.32
United Kingdom	50.55	782	3.07	79.41	17.52
Saudi Arabia	49.74	208	0.00	15.38	84.62

national emitter, has a larger Gini coefficient of 48.86 for 2,612 fossil-fuel plants (more than twice as many as China), 53.52 percent of which use natural gas as their primary fuel source.

India and Russia, both large developing nations, are ranked third and fourth, respectively, in national emissions and have Gini coefficients relatively similar in value to that of the United States but with a fraction of the number of plants. For India's 737 plants, 43.42 percent burn liquid fossil fuels, 37.72 percent burn coal, and 18.86 percent burn gas fossil fuels as their primary fuel. However, 65.97 percent of Russia's 529 plants burn gas fossil fuels as their primary fuel, with the remaining plants split between coal (19.28 percent) and liquid fossil fuels (14.75 percent).

The United Kingdom and Saudi Arabia are ranked ninth and tenth in national emissions and have Gini coefficients of 50.55 and 49.74, respectively. The majority of the United Kingdom's 782 fossil-fuel plants burn gas fossil fuels (79.41 percent) as their primary fuel, while the majority of Saudi Arabia's 208 fossil-fuel plants burn liquid fossil fuels (84.62 percent).

For all 161 nations in the study, the national-level Gini coefficients for disproportionality in plant-level emissions are moderately correlated with the number of fossil-fuel plants within a nation (the Pearson's correlation coefficient is .42). This result suggests that there is a tendency for facilities' emissions to be more dissimilar in countries with larger fleets of fossil-fuel power plants.[25]

Turning to other national characteristics known to drive emissions, the disproportionality Gini coefficients are correlated at .13 with the percentage of these nations' plants that primarily burn coal, .27 with the percentage of their plants that primarily burn gas fossil fuels, −.30 with the percentage of their plants that primarily burn liquid fossil fuels, .17 with gross domestic product per capita (measured in 2005 U.S. dollars), and, perhaps most importantly, .22 with national-level carbon emissions from fossil-fuel power plants.

MODELING THE EFFECT OF DISPROPORTIONALITY ON NATIONAL-LEVEL EMISSIONS: MULTIPLE REGRESSION ANALYSIS

Next, to evaluate whether disproportionalities in plant-level emissions are associated with these 161 nations' overall emissions, we conducted a multiple regression analysis to estimate the net effect of the disproportionality in power plant emissions within nations on national-level carbon emissions from fossil-fuel power plants for 2009, while also taking into account the effects of well-established human drivers of national-level emissions.[26] The dependent variable in the regression analysis is national-level carbon emissions from fossil-fuel power plants for 2009.

We included as additional independent variables national measures of population size, gross domestic product per capita (GDP per capita), and trade as a percentage of gross domestic product. We also included as additional independent variables the number of fossil-fuel power plants within a nation, whether or not a nation is located in a tropical climate (coded 1 if a country's predominant latitude is less than 30 degrees from the equator), and the average price of electricity for each nation (measured in U.S. dollars). Finally, we included measures of the percentages of a nation's fossil-fuel power plants whose primary fuels are coal, gas fossil fuels, and liquid fossil fuels.

To estimate the regression models, we used ordinary least squares (OLS) regression and robust regression, a combined approach suggested by researchers when analyzing cross-sectional data (data for one time point) for nations.[27] To be consistent with past cross-national research on the human drivers of GHG emissions,[28] prior to estimating the regression models, we converted all nonbinary variables into logarithmic form, leading to the estimation of elasticity coefficients. The interpretation of an elasticity coefficient is very straightforward: it is the estimated net percentage change in the dependent variable associated with a 1 percent increase in the independent variable. We provide additional information for the dependent variable, all additional independent variables, and the regression model techniques in the appendix to chapter 2.

We estimated three different OLS and robust regression models. The first includes only the disproportionality Gini coefficient, population size, and GDP per capita. The second model includes all independent variables, while the third model includes only the independent variables that have statistically significant effects on national-level carbon emissions from fossil-fuel power plants in the second model. Full information on the estimated regression models is provided in table A2.3 in the appendix to chapter 2.

The Results

The elasticity coefficients for the disproportionality Gini coefficients in the three OLS models are positive and statistically significant and range in value from 0.55 in the second model (includes all independent variables) to 0.70 in the third model (includes only the predictors in the second model that are statistically significant). Thus, based on these models, a 1 percent increase in disproportionality in plant-level carbon emissions leads to, at minimum, a 0.55 percent increase in national-level carbon emissions from fossil-fuel power plants.[29]

For the three robust regression models, the estimated effects (elasticity coefficients) for the Gini coefficients are also positive and statistically significant and range in value from 0.37 to 0.47. Robust regression is a modeling approach that downweights the potential influence of influential cases, oftentimes leading to the estimation of coefficients that are relatively smaller than those derived from OLS regression. These findings suggest that a 1 percent increase in disproportionality in plant-level emissions leads to at least a 0.37 percent increase in national-level carbon emissions from fossil-fuel power plants.

Turning briefly to the other independent variables—population size, GDP per capita, and the percentage of the nations' fossil-fuel power plants that use coal as a primary fuel source—all exhibit positive and statistically significant effects on national emissions, with only marginal differences across the OLS and robust regression models. The elasticity coefficient for price of electricity is negative and statistically significant, is slightly larger in the robust regression models than in the OLS models, and suggests that higher prices for electricity lead to reductions in emissions. The estimated effects of all other independent variables are not statistically significant. These results for the additional independent variables are generally consistent with past cross-national research on carbon emissions and other environmental outcomes.[30]

CONCLUSION

In this chapter, we have applied Freudenburg's disproportionality approach to fossil-fuel power plants in the majority of the world's nations. We find that countries vary significantly in their level of disproportionality and that all nations exhibit some level of disproportionality in power plant emissions. The ten nations with the highest levels of CO_2 emissions all have disproportionality Gini coefficients above the average for all nations, suggesting that disproportionality in plant-level emissions is especially problematic in nations with the largest overall carbon footprints. Further, our regression analysis suggests a positive association between national-level carbon emissions from fossil-fuel power plants and disproportionality in plant-level emissions within nations, even after accounting for the most well-established human drivers of national carbon emissions. In particular, the results suggest that a 1 percent increase in disproportionality in plant-level emissions leads to, at minimum, a 0.37 percent increase in national emissions from fossil-fuel power plants.

It is important to note that the data examined here are a decade old. In the time since they were collected, the Taichung power station has announced plans to build two new 800 MW units by 2021,[31] and Russian energy company Inter RAO has proposed to build a coal plant with a generating capacity 45 percent greater than that of the Taichung facility.[32] Adding this behemoth to the world's fleet of power plants would be like adding pollution-equivalents of Switzerland *and* either Estonia or Bolivia.[33] In the United States, a host of utilities plan to run hyperemitting coal plants for decades.[34] Duke Energy's Gibson plant in Indiana, which is the fourth-dirtiest power plant in the nation, was originally scheduled to close one of its generators in 2024 and two more in 2026. But Duke has moved both dates back by several years. Similarly, DTE Energy plans to extend the operation of its facility in Monroe, Michigan, the eighth-dirtiest plant in the United States, to 2040. At the same time that some utilities are holding onto their largest plants because they can produce more electricity at lower costs, they are taking steps toward closing some of their smaller ones. For example, Duke plans to retire seven of its smallest coal plants by 2024, while DTE intends to replace four of its smallest coal units with a larger plant that runs on natural gas and renewables.

These and other developments underscore an important policy implication of our results: to reduce the contributions of the electricity sector to overall GHG emissions in order to meet goals established in the recent Paris Agreement, some nations should consider reducing disproportionality among their fossil-fuel power plants by targeting a modest number of plants in the upper end of the distribution that burn fuels less efficiently. Policies that require or incentivize power plants in the upper tail of the distribution to produce electricity more efficiently could potentially go a long way toward reducing the sector's contributions to climate change.

However, just as deliberations over the Paris Agreement were marked by observations of differentiated responsibility, no disproportionality policy fits all. In countries where larger emitters are harder to regulate, pursuing other policy strategies, such as targeting a greater number of smaller emitters, may be more effective. Moreover, a nation could have a relatively low disproportionality Gini coefficient but still possess a substantially large carbon footprint if the majority of its power plants are burning large quantities of fossil fuels and inefficiently producing electricity. Put simply, if a country has only large, dirty power plants all emitting a similar amount, its low Gini coefficient would mask its overall impact on the climate. In any case, this research on disproportionality clearly suggests that policies targeting super polluters at the facility level should be considered alongside those that focus on sector-wide characteristics, economy-wide conditions, and other factors to reduce human-caused CO_2 emissions.

The Gini coefficients for disproportionality in plant-level CO_2 emissions do not tell the whole story. A Gini coefficient does provide a single snapshot of how emissions are distributed among fossil-fuel power plants in a given nation, but like most past studies, it does not by itself distinguish which individual power plants contribute the most carbon emissions, and it does not explain what combinations of facility-level characteristics and broader national conditions are likely to be associated with hyperpolluting power plants. In the next chapter, we present a theoretically integrative approach and use multiple research methods to address these sorts of important questions.

3

RECIPES FOR DISASTER

How Social Structures Interact to Make Environmentally Destructive Plants Even More So

In the last chapter, we demonstrated that a small set of power plants often accounts for a substantial share of a nation's electricity-based carbon dioxide (CO_2) emissions. Here we delve into that finding further by investigating the structural profiles of the world's most "villainous" polluters. Specifically, we address the following: Which types of fossil-fuel power plants generate the most carbon pollution, and what global, political, and organizational features do they have in common?

A large body of sociological research on the determinants of carbon pollution has focused on the effects of different international relationships on emission outcomes.[1] In particular, researchers are engaged in an ongoing debate over whether a nation's emissions are a function of its location in the core, semiperiphery, or periphery zone of the world-system[2] or of its embeddedness in world society, reflected in its participation in international environmental organizations.[3] Others have faulted these global accounts for neglecting middle-range factors.[4] According to them, pollution outcomes are primarily the result of national political-legal systems that regulate the power relations among business, the state, and the citizenry.

None of these approaches, however, has been used to study the organizations through which electricity and carbon pollution are actually produced—power plants—and the disparities in their propensity to pollute. In particular, researchers have yet to examine how global and political structures combine with the organizational properties of plants to generate emissions.[5] Climate-disrupting emissions are too "wicked" and complex a problem to be analyzed by one framework alone. To address this problem effectively, we need to better understand where and why these polluting activities occur. As it stands, not only have scholars made little progress integrating competing perspectives on carbon pollution, but also our understanding of the social structural profiles of hyperpolluting power plants and the mechanisms that possibly explain their environmental behavior remains limited.

We seek to remedy this situation by sketching what we label a *pathways to carbon pollution* framework. According to our framework, power plants are the chief sites through which various global, political, and organizational structures[6] interact[7] to generate multiple routes to carbon pollution. Some of these pathways, which we refer to as coercive and quiescent configurations, enhance plants' ability to externalize their pollution by neutralizing and manipulating potential sources of resistance. Others, which we call expropriative and inertial configurations, inhibit plants' ability to curb emissions by subjecting them to opportunistic behavior and forces of inertia.

To evaluate our framework, we used the data set introduced in previous chapters on the CO_2 emissions and structural attributes of nearly twenty thousand fossil-fuel power plants throughout the world. We also used multilevel regression techniques in conjunction with innovative fuzzy-set analytic methods that can identify which combinations of factors are associated with a given outcome. Using these data and methods, we determined whether power plants' high CO_2 emission rates are a function of coercive, quiescent, expropriative, and inertial configurations, which we describe in detail in later sections. We also tested these configurations' effects net of other controls and their impact on plants' emission levels.

BACKGROUND

Global Explanations of Carbon Pollution

Greenhouse gas emissions from the electricity sector are projected to double and perhaps even triple from 2010 baseline levels by 2050 unless steps are taken to decarbonize energy.[8] In light of these and other forecasts, social scientists have investigated the causes of cross-national differences in electricity-based CO_2 emission outcomes. Led by structural human ecologists,[9] these researchers initially emphasized the effects of population growth, wealth, and other indicators of national energy consumption.[10] More recently, scholars have redirected their attention to global structures that shape where and how energy is produced.[11] They disagree, though, over which of these structures are most consequential.

On the one hand, world-systems scholars argue that nations' positions in the global economic hierarchy lock them into different trajectories of fossil-fuel use.[12] From their perspective, all countries are organized into a stratified political-economic interstate system that is largely controlled by wealthy and geopolitically powerful "core" nations.[13] These core nations, which include the United States, Japan, the majority of the western European countries, and, most recently, the BRIC countries (Brazil, Russia, India, and China), substantially control trade and financial relationships with other nations. Core nations are also the predominant global producers (and consumers), extracting the basic resources they need for production (e.g., fossil fuels) from and exporting their waste to poorer peripheral nations. Semiperipheral nations are intermediaries between the core and periphery, typically undergoing rapid industrialization. It follows that as the global economy expands, the relationship between development and environmental harms will remain constant and possibly intensify in magnitude through time.[14] These strong and potentially intensifying relationships between environmental harms and development are largely driven by the structure of global production and trade networks as well as by foreign investment.[15]

On the other hand, world society scholars contend that a nation's willingness to counter carbon pollution depends on its cultural integration into a global proenvironmental order.[16] According to them, nation-states are nested in a global environmental regime comprised of shared norms and a variety of global institutions that compels them to recognize the importance of environmental sustainability to their long-term survival.[17] Scholars in this tradition have found that stronger ties to world society are associated with improved environmental outcomes at the national level, including reductions in carbon emissions as well as increased environmental concern at the individual level.[18] They suggest further that world society prods industry—in an effort to be seen as more legitimate in light of growing environmental concerns—to become more ecologically rational, i.e., to weigh the costs and benefits of environmental degradation and to strive to reduce externalities. Proponents of this perspective, then, expect industries located in nations that are more embedded in the global environmental regime, as indicated by their participation in environmental international nongovernmental organizations (EINGOs) and environmental treaties, will be more likely to reduce their pollution through various reforms and improvements.

Political Explanations of Carbon Pollution

Political institutions researchers note that while world society scholars and, to a lesser extent, world-systems researchers acknowledge that the nation-state is still relevant in an era of globalization, neither focuses on the "stuff of politics"[19]or the arrangements for resolving issues like energy-environment conflicts. Consequently, they overlook how different national political systems may determine carbon pollution.[20] According to these critics, the effects of global factors are overshadowed by the unity of the business class, the role of government regulation, and the public's awareness of environmental issues at the national level. Rachael Shwom shows, for example, that the results of efforts to make U.S. manufacturers improve the energy efficiency of their home appliance products have often been determined by the federal government's willingness to threaten regulation and its openness to pressure from business coalitions and energy advocates.[21] Harland Prechel and Alesha Istvan's study of the U.S. electrical industry likewise attests to how national political-legal arrangements condition the release of toxins.[22]

Especially germane to our worldwide study of electricity-based CO_2 emissions, Mario Bergara, Witold Henisz, and Pablo Spiller argue that because electric utilities require huge sunk investments, operate through economies of scale, and provide energy services to large populations, political actors have an incentive to behave opportunistically and expropriate utilities' assets.[23] This is less likely to happen, though, in nations where regulatory decision-making requires formal or informal government procedures, the judiciary has the tradition or authority to review administrative decisions, and the government's horizon is relatively long. These checks and balances provide safeguards against unilateral changes to the rules governing the electricity sector. They keep transaction costs down and give capitalists and public company managers fewer financial incentives to externalize their plants' pollution.

Organizational Explanations of Carbon Pollution

Organizational scholars argue that because global and political accounts of carbon pollution focus on cross-national differences in emissions, they obscure the fact that some facilities emit vastly more pollutants than others even after taking into account differences in their productivity. Such disproportionalities in plant-level carbon emissions were the focus of chapter 2.

The fact that the electricity sector is responsible for the lion's share of carbon pollution and the plants with the highest emission levels also tend to have higher than average emission rates might suggest that the decision of international bodies like the United Nations Framework Convention on Climate Change's Conference of the Parties to set targets for plants' carbon intensities is the most effective means to a low-carbon future.[24] Whether this is the optimal strategy is open to debate and will be scrutinized in the next chapter. However, it does raise this question: Why do some power plants have higher CO_2 emission rates than others?

Obviously, the types of inputs used by plants may be an important factor. In addition, plants that can produce more electricity may benefit from economies of scale that enable them to invest in better pollution equipment and practices. It is doubtful, however, that the environmental performance of power plants is solely a function of their inputs and potential outputs. Rather, as organizational scholars suggest, the structural properties of facilities themselves likely play causal roles.

For example, Charles Perrow[25] contends that size, the most studied variable in the literature on organizational structure,[26] is a major determinant of environmental degradation in modern society, including carbon pollution.[27] According to him and others, size concentrates power and thus enables corporations to maintain the status quo through their control over markets, ability to shape public policy, and framing of environmental issues.[28] In the case of electricity, plants with parent firms that own a sizable share of a country's existing production capacity are well positioned to leverage those forms of structural, instrumental, and discursive power to resist pressures to change and adopt more advanced pollution equipment.[29]

Similarly, organizational ecologists argue that the oldest members of organizational populations are especially subject to inertia.[30] Change is not only difficult and rare for these organizations but hazardous as well because their survival often depends on their predictability. Older organizations, therefore, not only have more vested interests in established pollution practices but also are prone to ossify.

Unfortunately, because they tend to view organizations as the "ultimate environmental destroyers,"[31] organizational scholars have been slow to acknowledge the possibility that the structural properties of organizations may combine with those of their external environments to shape emission outcomes, though that is beginning to change, as we note later. In that respect, they resemble world society, world-system, and political institutions researchers who treat their various structures as competing predictors of carbon pollution.

It follows from the preceding discussion that to advance our understanding of the structural determinants of electricity-based CO_2 emissions, we need an alternative conceptual framework that addresses how global, political, and organizational factors interact to shape power plants' emission outcomes. Toward that end, in the next section we draw on the literature on organizational configurations to sketch a theory of the pathways to carbon pollution.

THEORY: PATHWAYS TO CARBON POLLUTION

A configurational perspective suggests that to fully understand the impacts of organizations, we need to study the structures that differentiate them and the effects of those structures on outcomes like pollution.[32] These structures need not exist at the same level of analysis but can combine both within and across levels.[33] Indeed, it is through the discovery of these complementarities that researchers can begin to integrate insights from otherwise competing explanations of an organizational outcome, like plant-level emissions.

A configurational perspective conceives organizations as constellations of *interconnected* internal and external structures. In this view, organizations' structures are not entirely modular and thus should not be studied as individual predictors. Instead, organizations are made up of different bundles of structures, and scholars and researchers should, therefore, show how certain structural profiles are related to outcomes. So rather than asking whether age, size, location in the world economy, or civic engagement exerts the strongest independent effect on pollution outcomes, we need to ask how these factors combine to produce environmental degradation. A configurational perspective also rejects the notion of unifinality, which suggests that there is one optimal configuration leading to an outcome. Instead, it embraces the concept of equifinality, which states that two or more configurations can be equally effective in producing an outcome. It would suggest, therefore, there may be multiple and distinct combinations of internal and external factors leading to the same harmful pollution outcome.

Although prior research on carbon pollution has conceived global, political, and organizational conditions as alternative explanations, they are not fundamentally contradictory, since they highlight different structural properties. It is quite possible, therefore, that they work in concert to produce emission outcomes, as a configurational framework would expect. Indeed, environmental sociologists have begun exploring ways of integrating insights from different perspectives to provide more complete and nuanced understandings of the determinants of carbon pollution. For example, while world-system scholars continue to take issue with world society researchers' claim that worldwide norms are capable of offsetting the pollution created by global capitalism, some concede that nations that exclude transnationally organized civil society groups should have a more difficult time suppressing the environmental harms associated with trade regimes and global production networks than would nations with ties to these groups.[34] Similarly, some scholars who emphasize the agency of individual organizations and the effect their internal dynamics has on carbon emissions also acknowledge that the larger economic, cultural, and governance structures in which organizations are embedded can simultaneously influence pollution behavior.[35]

We extend these lines of inquiry by hypothesizing four ways that global, political, and organizational factors might interact to generate plant-level carbon pollution and suggest the mechanisms that explain their effects (as indicated by the labels we assign to them). Configurations that enhance plants' tendency to pollute come in two basic forms, which we label coercive and quiescent configurations. Those that inhibit plants' capacity to act in environmentally sustainable ways also come in two varieties, which we call expropriative and inertial configurations.

Coercive configurations resemble patterns that environmental inequality scholars suggest present polluters with the least resistance.[36] According to their research, eco-destructive facilities are often found in settings that have not only abundant market opportunities but also residents who are socially isolated and denied access to procedural equity.[37] Applied to our study of power plants' emissions, this would suggest that in core nations that are disengaged from global environmental norms and have governments lacking a system of political checks and balances (e.g., the United Arab Emirates, which according to Rob Clark and Jason Beckfield[38] has recently entered the core), utilities are able to engage in what Max Weber called coercive power[39] and impose their polluting activities on others.

Quiescent configurations mirror other scenarios suggested by environmental inequality scholars that keep citizens silent about pollution problems.[40] Under these scenarios, plants also seek out promising markets with residents who are socially isolated. But unlike the case with settings that offer the least resistance, the economic power of elites is decisive. That is, facilities owned by companies that control a large share of a market are able to forestall environmental grievances and mollify publics that depend on their products or services. It follows that the power plants most able to pollute by rendering the public compliant are those located in core nations that are disengaged from global environmental norms and have parent firms that dominate these nations' utility sectors. These plants exercise an influence akin to what Weber called manipulative power.[41]

Expropriative configurations correspond with circumstances that, according to transaction cost researchers, make utilities highly susceptible to opportunistic behavior by outside actors.[42] As discussed earlier, where there are few political constraints in the form of checks and balances, judiciary review, decentralized decision-making, etc., governments are apt to expropriate the assets of power stations for political gains. Although coercion and expropriation are related, the latter is understood here as differing in that it involves a government "taking" without adequate compensation. Large, well-established utilities are especially desirable targets for seizure because they service numerous citizens and have more accumulated knowledge about electricity production. Because the threat of being expropriated increases transaction costs, power plants that are located in nations lacking a system of political checks and balances, that have parent firms that dominate these nations' utility sectors, and that are old should be especially inclined to pollute.

Inertial configurations relate to conditions that world-system scholars and organizational ecologists suggest induce ossification. World-system scholars[43] contend that while core countries initially experience phases of growth and expansion, they often go through phases of relative decline and decay. Organizational ecologists[44] add that both larger and older organizations suffer from increasing calcification as a result of bureaucratization and other time-dependent processes. Moreover, both these types of organizations lose their ability to respond quickly or appropriately to changing conditions, and their technologies are prone to obsolescence. This suggests that power plants should have the most difficulty reducing their discharges when they possess these three inertial conditions simultaneously: they are located in a core nation; are part of a large, dominant parent firm; and are old.

Restated as formal hypotheses, we tested the following in the subsequent analyses:

1. There is no single structural pathway to carbon pollution. Instead, the power plants with the highest CO_2 emission rates are those that (a) are located in the world-system's core zone

and in nations that are disengaged from global environmental norms *and* that lack a system of political checks and balances (*coercive configurations*), (b) are located in the world-system's core zone *and* in nations that are disengaged from global environmental norms *and* are owned by dominant utilities (*quiescent configurations*), (c) are located in nations that lack a system of political checks and balances *and* are owned by dominant utilities *and* are old (*expropriative configurations*), and (d) are located in the world-system's core zone *and* are owned by dominant utilities *and* are old (*inertial configurations*).

2. These four pathways should significantly affect plants' emission rates net of other factors.
3. The underlying mechanisms that explain why plants on the four pathways have high emission rates are also relevant to their emission levels and, therefore, should have significant, positive net effects on both outcomes.

The appendix to chapter 3 (at the end of the book) provides in-depth descriptions of the data that we analyzed and the two methods, fuzzy-set qualitative comparative analysis (fsQCA) and multilevel regression, that we used to test our hypotheses.

RESULTS

fsQCA

As shown in table 3.1, we used fsQCA to test our first hypothesis that global, political, and organizational factors combine to create multiple pathways to power plants' emission rates. The table reports the final reduction set, and in keeping with our first hypothesis, it reveals there are multiple (five) configurations that produce high carbon intensities. Of the five configurations shown

TABLE 3.1 FINAL CONFIGURATIONS AND COVERAGE STATISTICS OF FSQCA ANALYSIS

Final Reduction Set[a]			
Set	**Raw Coverage**	**Unique Coverage**	**Solution Consistency**
Cnp	0.525	0.050	0.880
CnD	0.469	0.010	0.960
CnA	0.453	0.026	0.885
pDA	0.439	0.048	0.951
CDA	0.466	0.038	0.925

Total coverage = 0.739, Solution consistency = 0.830
[a]Configurations in bold were predicted.
C = core; N = normative engagement; P = political checks and balances;
D = dominant utility; A = plant age

here, four (designated in parentheses in the following paragraph) match our earlier descriptions of coercive, quiescent, expropriative, and inertial pathways.

When interpreting fsQCA results, it is important to bear in mind that no single attribute within a configuration can be interpreted outside the context of the other attributes. This is because, unlike regression techniques that abstract variables from the cases in which they exist, fsQCA treats individual cases as combinations of attributes. Also, in Table 3.1, upper case letters indicate that a condition is relatively present, whereas lower case letters indicate it is relatively absent.

Plants defined by the first configuration are facilities in core countries that are normatively disengaged and have few political checks and balances (coercive configuration). Plants defined by the second configuration are facilities owned by dominant utilities in core countries that are normatively disengaged (quiescent configuration). Plants defined by the third configuration, which we did not predict, are older facilities in core countries that are civically disengaged. Plants defined by the fourth configuration are older facilities owned by dominant utilities in countries with few political checks and balances (expropriative configuration). And plants defined by the fifth configuration are older facilities owned by dominant utilities in core countries (inertial configuration).[45]

Multilevel Regression

To assess the robustness of the configurations, we conducted multilevel regression analyses with various controls, as shown in table 3.2. With respect to the controls, the findings for model 1 show that plants tend to emit CO_2 at significantly higher rates when they rely primarily on coal, whereas, plants have lower rates when they rely primarily on gas; when they have more electrical capacity, which allows for economies of scale; when they are private and independently owned; when they are located in wealthier countries; and when a carbon tax is present. For model 2, we tested the effects of the configurations identified in the previous fsQCA analysis net of these controls. Here we see that the one configuration for which we did not develop a theoretically motivated configuration—old plants in normatively disengaged core countries—fails to exert a significant effect on emission rates beyond those of the controls. However, each of the other four that we did predict does have a significant, positive effect on plants' carbon intensities. The Bayesian information criterion (BIC) statistics indicate that the inclusion of the configurations provides a better model fit than a specification with just the controls.[46] As a final robustness check, we included previous emission rates as a control in model 3 and found that each of the significant effects of the four configurations still holds. Hence, the results of the regression analysis strongly support our

TABLE 3.2 REGRESSION ANALYSIS OF THE EFFECTS OF CONFIGURATIONS ON POWER PLANTS' CO_2 EMISSION RATES, 2009

	Model 1	Model 2	Model 3
Primarily coal	.451***	.449***	.114***
	(.016)	(.016)	(.013)
Primarily gas	-.147**	-.143***	-.117***
	(.011)	(.011)	(.009)
Electrical capacity	-.031***	-.039***	-.018***
	(.002)	(.002)	(.002)
Private ownership	-.060**	.009	.005
	(.015)	(.016)	(.013)
Population size	-.003	-.056***	-.014
	(.013)	(.017)	(.012)
Wealth	-.041***	-.065***	-.009
	(.014)	(.014)	(.009)
Price of electricity	-.066	-.038	-.037
	(.050)	(.045)	(.030)
Carbon tax	-.332**	-.192	-.010
	(.147)	(.155)	(.081)
Exports	.059	.001	-.006
	(.041)	(.043)	(.029)
Foreign direct investment	.001	-.017	-.003
	(.020)	(.019)	(.013)
Plants in normatively disengaged core countries with few political checks and balances (Cnp)		.338**	.143*
		(.120)	(.071)
Plants of dominant utilities in normatively disengaged core countries (CnD)		.748***	.409***
		(.081)	(.059)
Old plants in normatively disengaged core countries (CnA)		-.168	-.132
		(.136)	(.129)
Old plants of dominant utilities in countries with few political checks and balances (pDA)		.147**	.231***
		(.051)	(.042)
Old plants of dominant utilities in core countries (CDA)		.399***	.143***
		(.062)	(.051)
Prior CO_2 emission rate			.634***
			(.005)
Constant	6.715***	7.907***	2.641***
Random effects of countries (N = 148; x̄ obs. per group = 122)	.033***	.022***	.009***
Random effects of companies (N = 7,583; x̄ obs. per group = 2.5)	.139***	.128***	.047***
Residual variance	.201***	.202***	.109***
BIC	31165	29991	13808
N	19,525	19,525	19,525

Note: Standard errors are in parentheses.
$^{*}p < .05$
$^{**}p < .01$
$^{***}p < .001$ (two-tailed test)

second hypothesis that configurations of coercion, quiescence, expropriation, and inertia all significantly increase plants' emission rates net of other predictors.

Table 3.3 presents the test results for our third hypothesis: that these four configurations also exert a significant, positive effect on the absolute amount of CO_2 that plants emit. Here we replicated our regression analysis, but this time we substituted plants' emission levels as the dependent variable. A comparison of the BIC statistics for models 1 and 2 reveals again that the addition of the configurations provides a better model fit. More importantly, although this outcome measure and the previous one are only weakly correlated (.184), each of the same four configurations exerts a significant net effect on emission levels and continues to do so after another control for previous emission levels is added to the model. Hence, the results support our hypothesis that coercive, quiescent, expropriative, and inertial configurations increase not only power plants' emission rates but their emission levels as well. Figure 3.1 illustrates these four "recipes for disaster" using a star chart.[47]

DISCUSSION

Besides confirming our hypotheses about coercive, quiescent, expropriative, and inertial configurations, our results point to several ways research on carbon pollution might be advanced. First, researchers can use configurational analyses like the one we conducted to develop more nuanced theoretical arguments. Common explanations have asserted that particular structures like international trade regimes cause pollution and have accorded them a position of conceptual and causal priority. As a result, theorization about the different ways global, political,

TABLE 3.3 REGRESSION ANALYSIS OF THE EFFECTS OF CONFIGURATIONS ON POWER PLANTS' CO_2 EMISSION LEVELS, 2009

	Model 1	Model 2	Model 3
Primarily coal	.675***	.672***	.444***
	(.016)	(.017)	(.017)
Primarily gas	-.034**	-.037**	-.063***
	(.011)	(.011)	(.011)
Electrical capacity	.967***	.963***	.744***
	(.002)	(.002)	(.004)
Capacity utilization rate	.868***	.871***	.707***
	(.003)	(.004)	(.004)
Private ownership	-.042**	.003	-.019
	(.014)	(.015)	(.016)
Population size	-.007	-.058**	-.060**
	(.015)	(.020)	(.018)
Wealth	-.042*	-.065***	-.062***
	(.018)	(.016)	(.015)
Price of electricity	-.071	-.040	-.011
	(.060)	(.053)	(.032)
Carbon tax	-.391*	-.261	-.262
	(.188)	(.157)	(.136)
Exports	.047	.009	-.028
	(.049)	(.052)	(.046)
Foreign direct investment	.009	-.012	-.028
	(.023)	(.022)	(.020)
Plants in normatively disengaged core countries with few political checks and balances (Cnp)		.339**	.286**
		(.120)	(.107)
Plants of dominant utilities in normatively disengaged core countries (CnD)		.398***	.462***
		(.080)	(.073)
Old plants in normatively disengaged core countries (CnA)		-.114	-.247
		(.135)	(.235)
Old plants of dominant utilities in countries with few political checks and balances (pDA)		.099*	.066*
		(.048)	(.032)
Old plants of dominant utilities in core countries (CDA)		.326***	.291***
		(.060)	(.062)
Prior CO2 emission level			.234***
			(.004)
Constant	8.739***	9.874***	8.181***
Random effects of countries (N = 148; x̄ obs. per group = 122)	.058***	.037***	.028***
Random effects of companies (N = 7,583; x̄ obs. per group = 2.5)	.130***	.123***	.062***
Residual variance	.201***	.202***	.170***
BIC	29736	28693	19425
N	19,525	19,525	19,525

$^{*}p < .05$
$^{**}p < .01$
$^{***}p < .001$ (two-tailed test)

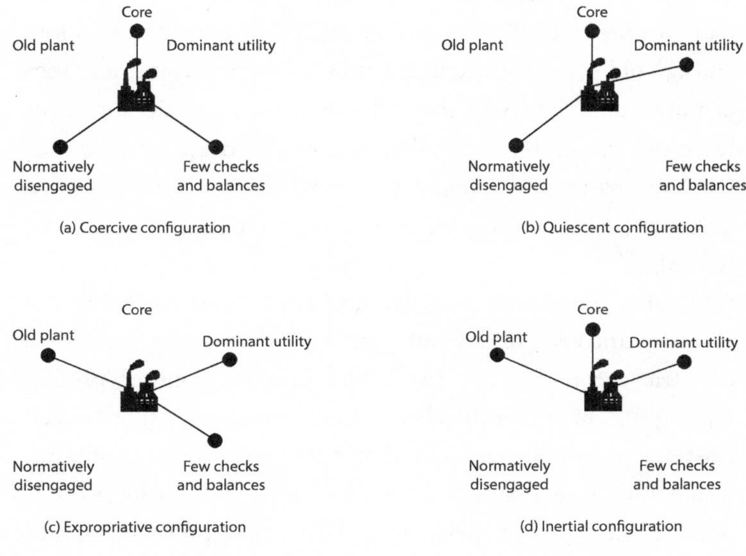

FIGURE 3.1 Star charts of extreme polluter configurations.

and organizational factors might interact to produce carbon pollution is underdeveloped. In contrast, our configurational framework shifts attention to how factors can jointly influence pollution outcomes and thus lays the conceptual groundwork for inquiries into the synergistic determinants of carbon emissions. It assumes there are multiple pathways to an outcome, each consisting of combinations of predictors. Changing any one predictor, therefore, will not necessarily affect the outcome because there are other pathways.

Researchers can also construct an empirical typology from the analysis above. Our findings suggest there are four distinct hyper-polluters: (1) plants in normatively disengaged core countries with few political checks and balances, (2) plants of dominant utilities in normatively disengaged core countries, (3) old plants of dominant utilities in countries with few political checks and balances, and (4) old plants of dominant utilities in core countries. They can use this typology to select the most causally relevant plants for more in-depth case studies and to determine whether the mechanisms that we suggest to explain their effects—coercive, quiescent, expropriative, and inertial—are accurate or need to be refined.

Finally, researchers could examine configurations within and across countries. For example, table 3.4 shows how individual countries score on the four configurations/polluter profiles highlighted here. Specifically, it lists the countries that have at least one plant that is more in than out of a particular configuration/profile (>.65) or that have a concrete, real-world example of a specific type of hyperpolluter. Using this information, they might compare the experiences of the two types of hyperpolluters found in China, which have what we call expropriative and inertial profiles, to get a better sense of how these mechanisms drive emissions in the world's most heavily polluting country. They might contrast the experiences of inertial plants found in traditional core countries (e.g., United States, United Kingdom) with those in countries that have recently entered that zone, including several Asian nations (China, India, Indonesia, Malaysia, Pakistan, Singapore, South Korea, and Thailand); in the "depoliticized" countries of eastern Europe (Bulgaria, Czech Republic, Hungary, Poland, Romania, and Russia); in nations in the Middle East (Egypt, Iran, Morocco, Saudi Arabia, and Turkey); and in countries with large economies in Latin America (Brazil and Mexico).[48] They might also investigate why particular nations (India, Indonesia, Iran, Pakistan, Saudi Arabia, Singapore, and United Arab Emirates) have all four types of hyperpolluters.

TABLE 3.4 CONFIGURATIONS/POLLUTER PROFILES BY COUNTRY[a]

Coercive	Quiescent	Expropriative	Inertial
India	Egypt	Afghanistan	Australia
Indonesia	India	Angola	Austria
Iran	Iran	Azerbaijan	Belgium
Pakistan	New Zealand	Bahrain	Brazil
Saudi Arabia	Pakistan	Bangladesh	Bulgaria
Singapore	Saudi Arabia	Belarus	Canada
United Arab Emirates	Singapore	Cambodia	China
	South Korea	Cameroon	Czech Republic
	Thailand	Central African Republic	Denmark
	United States	China	Egypt
	United Arab Emirates	Comoros	Finland
		Cuba	France
		Ecuador	Germany
		El Salvador	Greece
		Fiji	Hungary
		Guinea	India
		Haiti	Indonesia
		Honduras	Iran
		India	Ireland
		Indonesia	Italy
		Iran	Japan
		Iraq	Malaysia
		Jordan	Mexico
		Kuwait	Morocco
		Laos	Norway
		Liberia	Pakistan
		Libya	Poland
		Madagascar	Portugal
		Mexico	Romania
		Nicaragua	Russia
		North Korea	Saudi Arabia
		Pakistan	Singapore
		Peru	Spain
		Qatar	Sweden
		Rwanda	Switzerland
		Saudi Arabia	Thailand
		Senegal	Turkey
		Singapore	Ukraine
		Slovenia	United Arab Emirates
		Solomon Islands	United Kingdom
		Spain	United States
		Sri Lanka	
		Sudan	
		Taiwan	
		Tunisia	
		Turkmenistan	
		United Arab Emirates	
		Vietnam	
		Yemen	
		Zambia	

[a] A country is reported to have at least one power plant with a particular configuration/profile if the plant's maximum value on that profile/configuration is >.65. Because individual plants can have values that exceed this threshold for more than one of these sets, their countries may be classified as belonging to more than one of the categories listed above.

CONCLUSION

The study in this chapter advances research on carbon pollution by examining the synergistic effects of global, political, and organizational factors on power plants' CO_2 emissions. Using a unique worldwide data set, we found that none of these factors, by itself, is sufficient to explain plants' environmental behavior. Instead, plants' emission rates are best explained by distinct combinations of these factors, four of which continue to exert significant effects after controlling for other predictors and have similar effects on plants' emission levels as well.

The recognition that climate change and other environmental problems pose a threat to the sustainability of human society has given rise to a variety of sociological perspectives on environmental impacts,[49] but scholars have struggled to assimilate their ideas. One integrative approach is to test perspectives empirically in a common framework so as to determine which of their predictors has the strongest relative impact on aggregate pollution outcomes.[50] An alternative approach, which we are advocating, explores predictors' synergistic effects. This approach is more in keeping with the conclusion of the United Nations and others[51] that focusing on reducing the emissions of specific sectors or individual polluters is a more manageable strategy than targeting polluters across entire countries. And consistent with the notion that climate change is a "wicked problem" that often blends into other issues, our approach investigates the complex ways the causes of carbon pollution may interact.

This study also makes a number of contributions to research on the determinants of organizations' environmental performance. We advance research on the effects of external factors by examining how global economic and normative structures have differential effects on individual organizations within nations, an assumption often implied but rarely examined empirically. More specifically, we extend world-systems scholarship on the consequences of the core-periphery hierarchy for cross-national differences in carbon pollution and other environmental outcomes by examining the implications of this hierarchy for the environmental behavior of smaller social units (facilities) nested within it.[52] And we further world society research on how nations' commitments to global expectations and their actual environmental practices are sometimes decoupled by shifting attention to the sites where decoupling most likely occurs—power plants.[53]

In demonstrating that organizations' environmental performance is also conditioned by such external factors as national political institutions, we buttress the argument of other researchers that the state is a key strategic action field in which the rules for legitimate collective action are defined and enforced.[54] By the same token, we go beyond this research in advancing a strategy for studying the possible interactions among incumbents that wield disproportionate influence (e.g., members of the fossil-fuel industry), challengers that occupy less privileged positions (e.g., EINGOs), and governance units that more or less concentrate political power. Thus, our approach is consistent with the position of those who warn against making any one actor the central animating force in a field.[55]

In showing that carbon pollution is shaped by the larger economic, normative, and political environments in which power plants operate, we bolster the claim of open systems scholars[56] that an organization's behavior is conditioned by its external properties. At the same time, we improve on open system accounts by directing attention to those organizational actors that do the most environmental harm. Organizational scholars have been slow to study how and why particular kinds of power plants emit more carbon pollution than most, even though the issue of climate change will likely shape the relevance of future organizational studies.[57] Our configurational approach can also be extended to future studies of other types of organizations that may have disproportionate impacts on specific outcomes depending on their structures, such as universities and inequality.

And we improve on open system accounts by showing how we might analyze the interplay between external features of organizations. When organizational scholars studied the effects of external features in the past, they typically tried to statistically isolate the effects of those features of organizations. Or on the occasions when they examined how external features interact, they stopped short of determining which of the theoretically possible interactions are the most causally relevant. And despite the fact that some have endorsed the concept of equifinality—the idea that there are different causal pathways to the same outcome—they have relied solely on standard regression techniques that estimate a single path for all cases under examination.

In contrast, our configurational approach treats organizations holistically in the sense that each individual case is considered a complex entity. It also assumes that the attributes that causally define organizations fall into coherent patterns and, therefore, uses a methodology (fsQCA) that can identify which of the possible combinations of attributes are needed to explain a large fraction of outcomes.

Having suggested the strengths of our approach, it is important to acknowledge its limitations. First, theorization about and empirical analyses of the synergistic sources of carbon pollution are still in their infancy. In trying to make the case for a configurational model, we had few configurational predictions to build on and were, therefore, forced to develop hypotheses partly from scratch. Nonetheless, we believe that with more rough-cut slabs of findings like those presented here, the empirical foundations of an original configurational theory of carbon pollution can be slowly created. And as others become more familiar with fsQCA techniques, they will appreciate how they can provide the analytical legs on which such a theory needs to run.

Second, if this line of research is to be fruitful, additional theorization is needed to explain what generates configurations. We have suggested that processes having to do with coercion, quiescence, expropriation, and inertia may explain the observed effects of configurations. However, while several exogenous and endogenous forces seem capable of causing global, political, and organizational factors to cluster systematically, we have not explored them here. These range from environmental selection processes[58] and the diffusion of mimetic and normative strategies[59] to functional relationships among structures[60] and the replication of time-honored practices.[61] In addition to determining which of these forces help create clusters, scholars should investigate the stability of clusters over time, which would shed new light on the emergence and survival of particular organizational forms.

Third, as suggested earlier, additional research is needed to explain the effects of particular factors. As more nuanced worldwide data become available on power plants' ownership status and other corporate structures, researchers should determine if the effect of dominant utilities depends on, for example, whether they are owned by the state or private investors. Likewise, it will be important to elaborate on the effects of nations' political institutions to find out, for example, how differences in national climate change policies, market orientations, military investments, lobbying activities, entrepreneurship, and income inequality affect emission outcomes.[62] To better understand the effects of environmental norms, researchers should also examine the ability of EINGOs to construct collective action frames through the national media to delegitimize organizations that exploit natural resources.[63] Although collecting information on frames that target specific organizations can be time and cost prohibitive, with the advent of big data such tasks are now increasingly doable and could provide valuable insights into the impact of socially constructed landscapes on organizations' environmental behavior.

Fourth, while our study sheds light on the profiles of hyperpolluting power plants, it says nothing about those of green stars. Additional research is needed on these organizations in the electricity sector and other segments of the economy that have taken a low-carbon pathway. Using the configurational framework and strategy advocated here, scholars can begin to determine which constellations of global, political, and organizational properties distinguish these environmental leaders.

Fifth, research on hyperpolluting power plants needs to be extended beyond the period examined here when not only are the effects of climate change increasingly felt but also nationalist movements have grown. The unfolding environmental crisis in North and Southeast Asia has reinforced that region's predisposition toward top-down solutions. At the same time, more leaders of authoritarian populist movements in the European Union and United States have been elected to public office, threatening to unmoor those countries' checks and balances and, in the case of the United States, withdrawing from global climate change agreements. A key question for future studies, therefore, is what the impacts the different forms of centralized decision-making have on the emission rates and emission levels of hyperpolluters.

Having examined the social determinants of power plants' CO_2 emissions, in the next chapter we shift our analytical focus and begin to investigate their possible mitigation. Specifically, we conduct a modest test of the immodest claim that enhancing plants' efficiency is a surefire way to decrease their carbon releases.

We examine whether technical improvements in U.S. plants' heat-rate efficiencies reduce their carbon emission rates and levels or whether such efficiencies inadvertently increase their emission levels by stimulating overall electrical output. We also assess the impact of efficiency on the emissions of plants in other nations throughout the world as well as in the United States and the degree to which this impact depends on global and organizational factors.

4

A WIN-WIN SOLUTION?

The Paradoxical Effects of Efficiency on Plants' CO_2 Emissions

ENERGY efficiency has been heralded as the cheapest, fastest, and politically easiest way to reduce power plants' carbon dioxide (CO_2) pollution. But is it the "no regrets" strategy that advocates make it out to be? Or might it actually increase plants' CO_2 emissions under certain conditions?

In recent years, several leading agencies and think tanks have cited improvements in energy efficiency as the key to decreasing electricity consumption and its associated greenhouse gas (GHG) emissions. The Intergovernmental Panel on Climate Change (IPCC), International Energy Agency (IEA), International Renewable Energy Agency (IRENA), World Energy Council (WEC), U.S. Environmental Protection Agency (EPA), and Natural Resources Defense Council (NRDC) all claim that the cost of saving energy is now lower than new power generation and could be readily dropped further by simply adopting existing green technologies. They add that energy efficiency is more palatable to voters because, unlike other environmental measures such as capping emissions or putting a price on carbon, it requires no sacrifice on their part and thus is a "no-cost," "win-win" solution.

Consistent with these groups' recommendations, over three-fourths of the countries that ratified the Paris Agreement and submitted nationally determined contributions have included provisions to enhance energy efficiency. In the United States, where political liberals and conservatives strongly differ over whether to support the Paris Agreement, both nonetheless advocate efficiency as a regulatory tool. Under the Obama administration's proposed Clean Power Plan, which was to serve as the central legal mechanism for meeting U.S. commitments made under the Paris Agreement, states must reduce their CO_2 emissions by choosing from several best systems of emission reductions, including improving individual plants' heat-rate efficiency (the amount of heat lost in the process of generating 1 kilowatt-hour [kWh] of electricity). Under the Trump administration's replacement for the Clean Power Plan, called the Affordable Clean Energy Rule, states are not subject to the Paris Agreement, and fossil-fuel plants are obligated only to convert coal, gas, or oil to electricity more efficiently. As noted in chapter 1, major polluters themselves like the Taichung power plant have also espoused the idea that efficiency is a simple fix to emission reductions.

Despite the wide acceptance of the efficiency approach, some have questioned its environmental benefits, claiming it could actually lead to increases in CO_2 emissions. Building on William Stanley Jevons's[1] classic argument that increasing the efficiency of coal use further diminishes its supply by ironically increasing demand for this resource through its "very economy," skeptics contend that efficiency encourages greater consumption of fossil fuels—a phenomenon called *rebound*, which, in its extreme form termed *backfire*, does not just reduce but also nullifies or reverses gains in efficiency. From their perspective, such backfires not only are inevitable but also strengthen the link between energy consumption and carbon pollution. This reasoning is consistent with sociological literatures on the modern world-system,[2] treadmill of production,[3] and structural human ecology,[4] which all suggest efficiency encourages further economic expansion, which, in turn, exacerbates carbon pollution.

Efficiency supporters dispute these skeptics' assertions, which they contend are based on faulty conclusions. They argue that as societies remove market barriers preventing the adoption of existing technologies that can extract vastly more economic benefit from a unit of energy, these efficient devices will enable economies to both expand energy usage and "decouple" it from the release of GHGs. Scholars in industrial ecology,[5] ecological modernization,[6] and the environmental Kuznets curve tradition[7] espouse a similar logic that energy efficiency allows societies to reduce their environmental impact without compromising their prosperity.

To date, little progress has been made in reconciling these opposing views because their proponents have yet to analyze the real-world relationship between power plants' efficiency heat rates and their CO_2 emissions. Nor have they examined which internal and external features of plants make them more or less susceptible to emission rebounds and backfires. In large part, this is because investigators have lacked access to systematic data on individual plants' heat rates and carbon discharges, and they have yet to use a conceptual framework attuned to the internal and external properties of power plants.

To remedy this situation, we conducted a modest test of the immodest claims made by efficiency supporters and critics, the results of which are presented in the chapter. Using the EPA's Greenhouse Gas Reporting Program (GHGRP) data, we first assessed the effect of efficiency on the reported rates and levels at which individual U.S. power plants emit CO_2, while accounting for the impacts of other relevant factors. We then broadened our analysis with the data set used in chapters 2 and 3, which contains nearly every fossil-fuel plant in the world, to examine the pollution impact of efficiency and the conditions under which it varies. Our hypotheses were informed by Max Weber's influential writings on the causes and consequences of efficiency in modern society.

Both sides of the efficiency/emissions debate frequently invoke Weber to justify their position. Supporters cite his statement about the "technical superiority" of efficiency, while critics quote his writings about efficiency's contradictory and destructive effects. However, as other scholars have recently argued, a closer reading of Weber's work reveals he had a more nuanced understanding of efficiency that suggests its effects are contingent on the microlevel attributes of actors as well as the macrolevel institutional settings in which they are embedded. Expanding on this logic, we developed and tested a series of hypotheses that start to reconcile the opposing claims of efficiency optimists and pessimists.

BACKGROUND

The electricity sector has the greatest potential for decarbonization of any sector.[8] To begin to realize that potential, experts have recommended a variety of steps that power plants can take to improve their efficiency, or generate the same amount of electricity from less fossil fuel, ranging from turbine upgrades, equipment refurbishment, condenser optimization, and boiler system improvements to better operating and management schedules and conversion from subcritical steam systems to supercritical or advanced ultracritical steam systems.[9] Agencies and consulting firms estimate that for every 1 percent improvement in plants' efficiency induced by these technologies, plants' emission rates (carbon pollutants emitted per unit of electrical output) could be decreased between 0.6 and 3 percent.[10] And if plants had incorporated all existing green technologies, efficiency measures alone could have reduced U.S. and global electricity-based emission levels (total pounds of CO_2 released) in 2020 by as much as 24 percent and 33 percent, respectively, compared to 2005 levels.[11]

Efficiency advocates typically base their optimistic estimates on the assumption that energy consumption will remain constant and aggregate gains in energy efficiency have a direct effect on GHG emissions. However, others have challenged this assumption, arguing it ignores the potential increases in energy consumption that are known to result from below-cost efficiency improvements[12] and can indirectly cause CO_2 emissions to rise. They observe, for instance, that efficiency improvements in refrigerators and air conditioners have driven down their costs of operation to such an extent that nearly all U.S. homes now own refrigerators and air conditioners, and oftentimes own multiple units, and use roughly as much electricity on these items as before improvements were made. They note further that in many countries, aggregate CO_2 emissions per dollar of gross domestic product, or GDP (a common proxy for carbon efficiency), and aggregate CO_2 emission rates have both steadily decreased,[13] while aggregate CO_2 emission levels have moved in the opposite direction,[14] indicative of emission backfires. Efficiency skeptics conclude that unless caps and price controls are in place to limit and discourage the use of fossil fuels, plants will inevitably use more fossil fuels as they become more affordable due to efficiency gains, ultimately causing the total amount of carbon emissions to escalate.

In response, advocates have argued that efficiency does increase productivity and should improve economic growth, but they steadfastly disagree with skeptics' suggestion that this necessarily results in emission backfires. They argue that whatever relationship exists between efficiency and emission increases is an artifact of other factors, such as wealth, inequality, and population. They concede that emission rebounds can occur, compromising the mitigating impact of efficiency on carbon pollution, but insist that rebounds are small. They cite research showing, for example, that direct rebounds for consumer energy services, especially in wealthier countries, are fairly modest, typically eroding between 10 and 30 percent of the energy savings.[15] To the extent rebounds happen, they are said to be due to an "energy-efficiency gap" caused by market barriers that prevent actors from adopting the best available green technologies. Advocates estimate that only a third of the electricity sector's efficiency potential is currently being realized. As market barriers are removed over time, they contend, efficiency will eventually reduce emissions to near zero.

While ideological differences may partly explain why this controversy has continued without any sign of resolution, a more plausible reason is that there has been a lack of systematic data needed to analyze the empirical relationship between individual power plants' heat rates and their emission outcomes. This has allowed both sides to assert their relatively simplistic assumptions without testing them. It also helps explain why studies by each side have relied heavily on simulations, forecasts, scenarios, and other predictive and descriptive techniques prized for their elegance and parsimony. A more realist or "dirty hands" approach,[16] capable of refining our understanding of emission rebounds and tailoring efficiency measures, would use explanatory methods suited for assessing theoretically informed causal relationships.

Because energy production accounts for an increasingly smaller percentage of most economies today, in large part because of advances in efficiency, many experts also now feel it is appropriate to study rebounds on the opposite end of the supply to demand (or production to consumption) chain. In fact, the vast majority of studies on the subject have focused on the direct rebounds among end-use consumers of energy and the environmental impacts of their more efficient vehicles, appliances, and other goods. The energy sector's changing relative size notwithstanding, the upstream producers of energy—power plants—still remain the world's most concentrated source of anthropogenic GHGs. And the electric utilities sector is the most responsive (*elastic*) to changes in energy prices and consumer demand as well as the most able to substitute a cheaper energy input for others, suggesting that improvements in energy efficiency in this sector are especially likely to generate energy rebound effects that could result in higher emission levels.

A recent study by Amelia Keyes and colleagues comparing the environmental consequences of President Barack Obama's Clean Power Plan and President Donald Trump's Affordable Clean Energy Rule[17] perhaps comes closest to redirecting attention back to power plants and the effects of their heat-rate efficiencies on emission outcomes. Using ICF International's Integrated Planning Model, the authors predict that the former, a "beyond the source" approach that allowed plants to reduce their emissions through on-site heat-rate efficiency improvements as well as generation shifting, renewable energy capacity increases, and other measures, would not have generated rebound effects. The latter, an "inside the fence line" regulation, however, abolishes all of the options available to plants under the Clean Power Plan except for the one preferred by the fossil-fuel industry—heat-rate efficiency improvements—which does not require transitioning to renewable energy sources. According to Keyes and colleagues, this regulatory approach will produce rebounds at state, regional, and national levels by allowing polluting coal plants to run much longer and well past their scheduled retirements.

Although this study and others like it advance our understanding of emission rebounds associated with the electric utilities sector, they have several shortcomings. First, because they simulate aggregate emission outcomes, they gloss over an issue discussed in chapter 2—the disproportionate amount of carbon released by some power plants relative to others. Second, they suggest that under certain policy scenarios, plants are apt to substitute cheaper and more carbon-intensive energy inputs, causing emission backfires to occur, but they fail to investigate which types of plants are more susceptible to such backfires in the first place. To identify the mechanisms that differentiate plants that do and do not increase their emission levels as a result of higher heat-rate efficiencies, researchers need to examine not only variations in individual plants' emissions but also properties of the plants themselves or how they are organized internally. Finally, and related to the last point, studies that focus solely on rebounds within a major polluting nation do not consider how plants are organized externally. That is, they neither account for the economic relationships that exist between a plant's country and others nor situate a plant's country in larger normative systems that shape how countries understand and respond to the global problem of climate change. This situation remains despite the fact that the success of the Paris Agreement and other treaties is, as many experts now recognize, greatly influenced by the interdependence of national economies and nongovernmental organizations' ability to frame and prioritize climate issues.

THEORY

Many sociologists, beginning with Weber, have argued that efficiency is a hallmark of modern times. According to them, as societies develop, they increasingly dispense with reasoning based on tradition and employ instead an instrumental form of rationality that seeks out the best means to desired ends. While efficiency creates numerous conveniences and greatly expands the supply of goods and services, it is also susceptible to destructive contradictions or, in Weber's terms, formal rationality (aimed at choosing the technically appropriate means) does not guarantee material rationality (intended to achieve the original value-based goal). Where a group is able to concentrate power, it may strategically use efficiency to protect and advance its interests at the expense of others. And to the extent that a powerful few can entrench or prioritize efficiency in public policy, efficiency can become an end in itself, causing goal displacement.

Inspired by Weber's theory of bureaucracies as autonomous, rational entities and his later writings about context-bound rationality, later generations of scholars have also reconceptualized modern organizations as closed and open systems. They suggest that contradictions may stem from the microlevel characteristics of the corporate actors themselves as well as their surrounding macrolevel institutional environments that promote practices and logics at odds with the aims of efficiency. For example, critics of bureaucracies, organizational ecologists, and other researchers who focus on organizations' internal properties argue that as organizations grow in size, they are increasingly reluctant to adopt practices or technologies with which they are less familiar,[18] including environmentally friendly ones, because their survival depends on their predictability. Similarly, as organizations become more established, they struggle to innovate due to growing inertia and senescence.[19]

Scholars more attuned to external forces suggest economic arrangements may partially dictate how organizations behave. For instance, proponents of the world-systems perspective contend countries are organized into an unequal interstate system that is largely controlled by a set of wealthy and geopolitically powerful "core" nations.[20] Not only do these nations dominate trade and financial relationships, but also their dependence on basic energy resources like fossil fuels for continued growth locks their businesses into environmentally destructive trajectories,[21] and their large and powerful militaries consume enormous amounts of fossil fuels and other nonrenewable resources.[22]

Cultural regimes may also determine the relative importance of issues and the types of normative pressures exerted on organizations. World society scholars, for example, contend organizations located in nations that are more embedded in the global environmental regime, as indicated by their participation in environmental international nongovernmental organizations (EINGOs), are more likely to adhere to global norms and recognize the importance of environmental sustainability.[23] However, to the extent EINGOs and associated reforms are decoupled from intended outcomes, organizations located in nations that participate in these regimes may use EINGOs as window dressing[24] to divert attention from these organizations' poor environmental behavior.

It follows that contradictions in the form of emission rebounds are especially likely among plants in countries like the United States, where the fossil-fuel industry has been especially powerful and aggressively promoted efficiency in ways that safeguard its interests. For plants more generally, those that are larger and older should have less flexibility and incentive to make efficiency improvements to achieve new environmental ends. To the extent these internal properties facilitate improvements in plants' efficiency, they will serve to increase electrical output only, causing emissions also to rise. Due to pressures to increase productivity at any cost, plants located in core nations within the global economy should have more difficulty translating efficiency gains into emission reductions. And to the degree a nation's ties to the global environmental regime function as an environmental façade, participation in EINGOs may enable plants to make efficiency improvements that exacerbate emissions.

In the next section, we test these hypotheses using samples of power plants in the United States and across the world. Details about these data, measures, and the methods used are provided in the appendix to chapter 4 (at the back of the book).

ANALYSIS

U.S. Sample of Power Plants

Figures 4.1 and 4.2 show the predicted effects of U.S. plants' heat rates on their emission rates and logged emission levels for 2010 using a bivariate regression (these data are logged due to their skewed distribution). Heat rate is the amount of energy used to generate 1 kWh of electricity, expressed in British thermal units per net kWh produced. Plants that score higher on this indicator do a less efficient job of capturing heat energy, and vice versa. Figure 4.1 shows that there is a positive relationship between heat and emission rates, indicating that the less efficient plants are at preventing heat losses, the less effective they are at preventing carbon releases per unit of electricity generated. Figure 4.2 reveals that heat rate and emission level have an opposite or negative relationship. This finding is in keeping with the skeptics' argument that although improvements in plants' efficiency can reduce the rate at which they pollute, they may actually worsen their overall volume of pollutants.

To determine whether the bivariate relationships observed in figures 4.1 and 4.2 can be explained by other relevant factors, we conducted multiple regression analyses of the determinants of plants' emission rates and levels for 2010. Specifically, in models 1–3 in table 4.1, we assessed the determinants of plants' CO_2 emission rates using a cumulative model that starts with a baseline equation containing control variables (model 1) and adds an

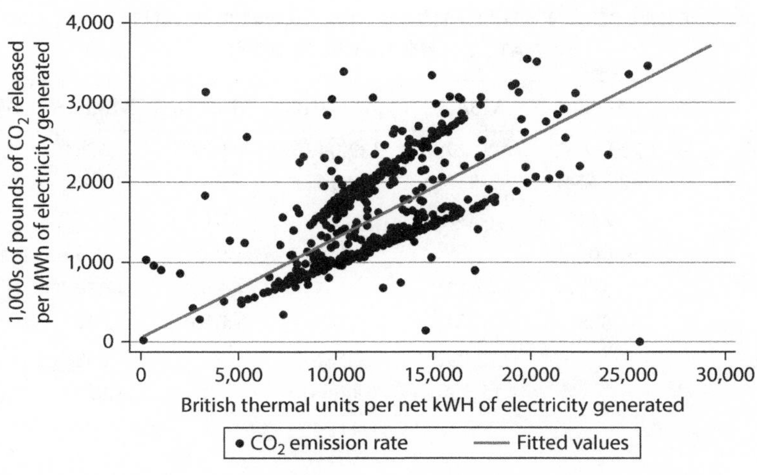

FIGURE 4.1 Predicted effect of heat rate on CO_2 emission rate.

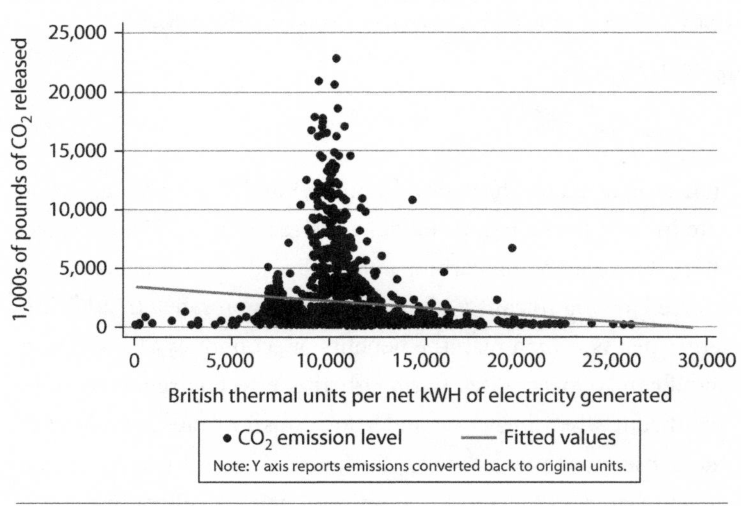

FIGURE 4.2 Predicted effect of heat rate on logged CO_2 emission level.

TABLE 4.1 REGRESSION ANALYSIS OF U.S. POWER PLANTS' CO_2 EMISSION RATES AND LEVELS, 2009

	Model 1	Model 2	Model 3	Model 4	Model 5	Model 6
Coal fuel	729.566**	629.298**	322.047**	1.639**	1.686**	.489**
(1 = yes)	(88.464)	(52.581)	(54.142)	(.211)	(.196)	(.152)
Size	-.244**	-.041	-.016	.020**	.010**	.004**
	(.063)	(.039)	(.033)	(.002)	(.001)	(.001)
Fuel price	-.093	-.612**	-.395*	-.030**	-.030**	-.011**
	(.338)	(.201)	(.173)	(.010)	(.010)	(.005)
Heat rate		113.090**	82.586**		-.148**	-.086**
		(5.630)	(5.664)		(.021)	(.015)
Emissions rate			.441***			
in 2005			(.044)			
Emissions level						.670**
in 2005						(.040)
Constant	1,459.911	263.258	-176.614	12.31	14.07	5.071
R^2	.232	.654	.767	.559	.613	.832
N	1,129	1,129	1,129	1,129	1,129	1,129

Notes: Regression coefficients are unstandardized; standard errors are in parentheses.
$^{*}p < .05$
$^{**}p < .01$ (two-tailed test)

indicator of plants' heat rate (model 2) and their CO_2 emission rate in 2005 (model 3). In model 1, we see, as would be expected, that plants relying on carbon-intensive coal as their primary fuel source have significantly higher CO_2 emission rates. In addition, larger plants, which reap the benefits of economies of scale, have significantly lower rates (although this effect is rendered non-significant when heat rate and 2005 emission rate are added in the subsequent two models). Fuel prices have a nonsignificant effect in model 1 but exert a significant, negative effect in the two models where heat rate and prior emission rate are included, suggesting that higher fuel prices force plants to economize and thus reduce emission rates after accounting for these two other factors. Model 2 reveals, as anticipated, that net of the controls, the higher a plant's heat rate, the higher its emission rate. As model 3 shows, this is true after controlling for plants' prior emission rates in 2005 as well.

In models 4–6, we used the same modeling strategy to assess the determinants of plants' emission levels for 2010. Here again coal fuel has a significant, positive effect across all three models. The same is true of size, which suggests that as a plant's capacity for generating electricity increases, so, too, does its potential for releasing greater volumes of carbon dioxide. As was the case with emission rates, higher fuel costs significantly decrease emission levels. Of special significance and consistent with the arguments of critics of energy efficiency as a strategy for reducing carbon emissions, model 5 reveals that net of controls, heat rate has a significant, inverse relationship with emission levels. This indicates that as plants become more efficient at preventing heat losses, their total emissions tend to grow. As model 6 reports, the effect of the heat rate on emission levels remains significant even after controlling for prior emission levels in 2005.

International Sample of Power Plants

Next, we expanded our analysis to include the world's entire fossil-fueled power fleet. Unlike the case with our U.S. data set, efficiency is measured in our international data file as thermal efficiency, or the total energy produced as a percentage of the heat energy generated. It is essentially the inverse of heat rate. Hence, plants that score higher on this indicator do a *more* efficient job of capturing heat energy. Keep this in mind when comparing the directional effects of heat rate discussed earlier and those of thermal efficiency reported in this section.

Figure 4.3 summarizes the correlation between plants' thermal efficiencies and CO_2 emissions for each country for 2009. Thermal efficiency (the inverse of a plant's heat rate) has a negative correlation with emissions, which energy efficiency proponents would expect, in countries like Russia (–.087), which has the fourth-largest amount of electricity-based CO_2 emissions. In contrast, countries like China (.073), the United States (.012), and India (.103), which have the first-, second-, and third-largest amounts of total emissions, respectively, exhibit positive correlations, consistent with a rebound effect. The correlations for these four countries, however, are all fairly weak. We also see a significant degree of variation across developing countries; many nations in Latin America exhibit direct rebounds, whereas most nations in sub-Saharan Africa do not.

Table 4.2 presents a multilevel regression analysis of the determinants of emission levels for 2009. Here we tested a model containing all of our key independent variables and controls (model 1) and then added a term for the interaction between thermal efficiency and each of our key independent variables in the subsequent four models. The first model reveals that plants' thermal efficiency has no significant net effect on emissions. In contrast, each of the other key independent variables is significantly related to plant-level emissions. Older and larger plants and those located in a core or semiperiphery nation (compared to a peripheral nation) emit larger amounts of emissions, whereas plants in countries with more EINGOs and higher electricity prices emit less. Plants that rely primarily on coal, fully utilize

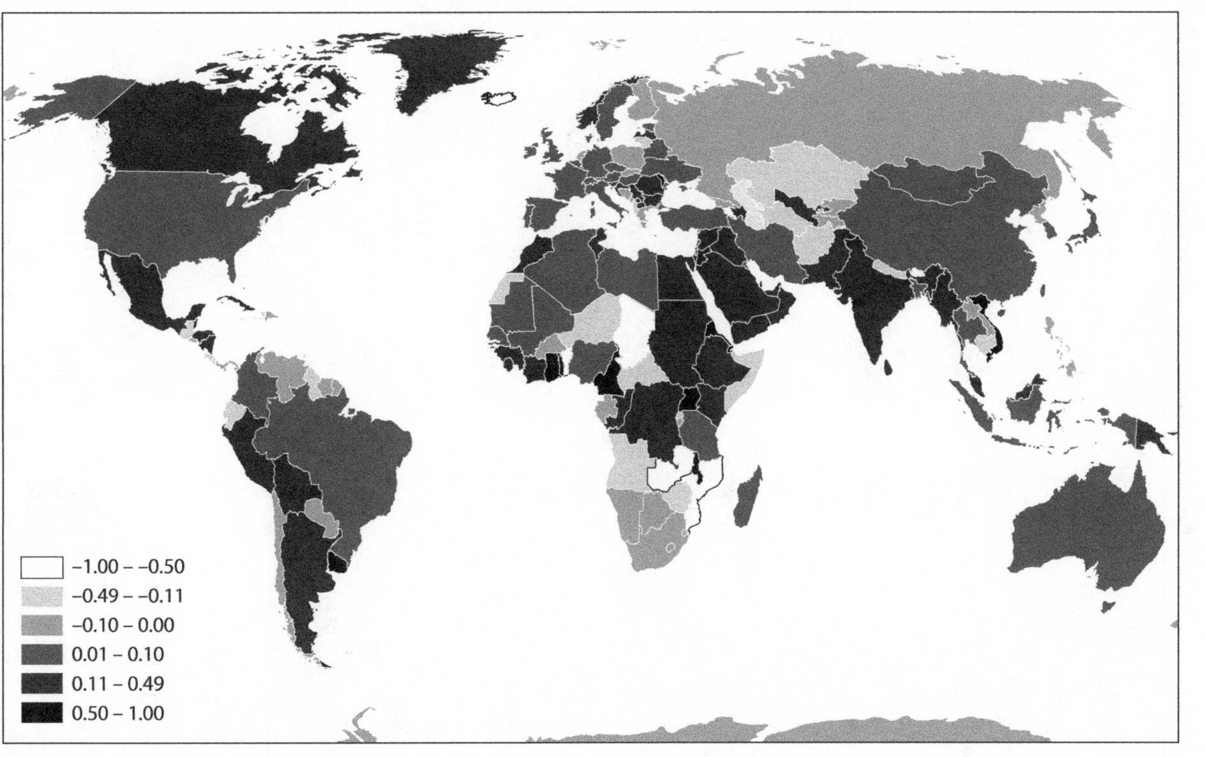

FIGURE 4.3 Correlation between plant-level CO_2 emissions and thermal efficiency by country, 2009.

TABLE 4.2 REGRESSION ANALYSIS OF THE DETERMINANTS OF POWER PLANTS' CO_2 EMISSIONS, 2009

	Model 1	Model 2	Model 3	Model 4	Model 5
Efficiency					
Thermal efficiency	-.001	.003	-.008**	-.029*	-.073**
	(.001)	(.003)	(.002)	(.012)	(.017)
Organizational					
Plant age	.018**	.017**	.018**	.018**	.018**
	(.001)	(.001)	(.001)	(.001)	(.001)
Plant size	.002**	.002**	.003**	.002**	.002**
	(.000)	(.000)	(.000)	(.000)	(.000)
World-System					
Core	1.651**	1.671**	1.656**	2.014**	1.671**
	(.211)	(.212)	(.212)	(.254)	(.212)
Semiperiphery	.827**	.827**	.812**	.818**	.806**
	(.182)	(.183)	(.182)	(.183)	(.182)
World Society					
EINGOs	-.787**	-.788**	-.785**	-.778**	-.534**
	(.091)	(.091)	(.091)	(.091)	(.108)
Controls					
Primarily coal	1.651**	1.653**	1.666**	1.650**	1.639**
	(.058)	(.058)	(.058)	(.058)	(.058)
Population	.121**	.122**	.125**	.122**	.120**
	(.033)	(.033)	(.033)	(.033)	(.033)
Capacity utilization rate	3.520**	3.516**	3.483**	3.526**	3.543**
	(.078)	(.078)	(.078)	(.078)	(.078)
Wealth	-.036	-.036	-.038	-.041	-.043
	(.040)	(.041)	(.041)	(.041)	(.041)
Price of electricity	-2.105**	-2.110**	-2.141**	-2.068**	-2.070**
	(.354)	(.355)	(.355)	(.355)	(.355)
Thermal efficiency * Plant age		-.001			
		(.001)			
Thermal efficiency * Plant size			.004***		
			(.001)		
Thermal efficiency * Core				.030**	
				(.012)	
Thermal efficiency * EINGOs					.021**
					(.005)
Constant	6.642**	6.674**	6.614**	6.312**	5.852*
R^2	.560	.560	.565	.564	.566
N	19,525	19,525	19,525	19,525	19,525

Note: Standard errors are in parentheses.
**p* < .05
***p* < .01 (two-tailed test)

their capacity, and are in highly populated countries have higher emission levels. National wealth has a nonsignificant relationship with plant-level emissions.

Model 2, in which thermal efficiency interacts with plant age, reveals that more efficient plants are no more likely to emit significantly greater amounts of CO_2 if they are old. However, the next three models show that more efficient plants emit at significantly higher levels when they are large (model 3), located in core countries (model 4), and situated in countries with more ties to the global environmental regime (model 5). (In unreported analyses, we found that semiperiphery location does not have significant interactions with thermal efficiency.) Model 5 suggests that while EINGOs directly reduce plants' emissions, they also decouple efficiency and carbon pollution, causing plants to increase their output and, in turn, their emissions as they become more efficient. Why the presence of more EINGOs might cause rebound effects is not immediately clear, but it is consistent with the argument of other sociologists that the expansion of civil society involvement in climate issues can sometimes have unintended consequences, including the disenfranchisement or cooptation of EINGOs.[25]

Table 4.3 replicates the same analyses but includes a measure of prior emission levels in 2004. Model 6 reveals that thermal efficiency directly reduces emission levels over time (2004–2009), as efficiency advocates would expect. The four remaining models also reveal that the effect of thermal efficiency varies depending on the organizational and global characteristics of plants. Specifically, more efficient plants emit at significantly higher levels when they are older (model 7), larger (model 8), in core nations (model 9), and exposed to more EINGOs (model 10). (Note that the effects of world-system position in these models are the opposite of what we observe in table 4.2 and that national-level wealth now has a significant, negative effect, all of which might partly

TABLE 4.3 LAGGED DEPENDENT-VARIABLE REGRESSION ANALYSIS OF THE DETERMINANTS OF POWER PLANTS' CO_2 EMISSIONS, 2009

	Model 6	Model 7	Model 8	Model 9	Model 10
Efficiency					
Thermal efficiency	–.025**	–.020**	–.009**	–.078**	–.132**
	(.007)	(.003)	(.001)	(.009)	(.013)
Organizational					
Plant age	.006**	.011**	.006**	.006**	.006**
	(.001)	(.001)	(.001)	(.001)	(.001)
Plant size	.040**	.040**	.007**	.040**	.049**
	(.002)	(.002)	(.001)	(.002)	(.002)
World-System					
Core	–.731**	–.729**	–.744**	.223	–.679**
	(.122)	(.122)	(.122)	(.166)	(.121)
Semiperiphery	–.887**	–.881**	–.906**	–.849**	–.857**
	(.108)	(.107)	(.107)	(.107)	(.107)
World Society					
EINGOs	.038	.043	.039	.057	.484**
	(.052)	(.053)	(.053)	(.052)	(.069)
Controls					
Primarily coal	.132**	.121**	.143**	.129**	.118**
	(.033)	(.033)	(.033)	(.033)	(.033)
Population	.023	.022	.025	.023	.022
	(.018)	(.018)	(.018)	(.018)	(.018)
Capacity utilization rate	1.347**	1.373**	1.315**	1.363**	1.389**
	(.046)	(.046)	(.046)	(.046)	(.046)
Wealth	–.066**	–.065**	–.069**	–.075**	–.073**
	(.022)	(.023)	(.023)	(.023)	(.023)
Price of electricity	.403	.438	.364	.501	.454
	(.297)	(.297)	(.298)	(.298)	(.297)
Lagged CO_2 emissions	.789**	.791**	.789**	.790**	.789**
	(.004)	(.004)	(.005)	(.005)	(.005)
Thermal efficiency * Plant age		.004**			
		(.001)			
Thermal efficiency * Plant size			.030**		
			(.005)		
Thermal efficiency * Core				.078**	
				(.009)	
Thermal efficiency * EINGOs					.038**
					(.004)
Constant	2.008**	6.674**	2.004**	1.048**	.852
R^2	.919	.922	.923	.923	.923
N	19,525	19,525	19,525	19,525	19,525

Note: Standard errors are in parentheses.

*$p < .05$

**$p < .01$ (two-tailed test)

reflect the global recession that occurred between 2004 and 2009. Also, the effects of the price of electricity and population are nonsignificant in the models that control for prior emission levels.)

DISCUSSION AND CONCLUSION

The results for our analysis of U.S. power plants are striking. More efficient fossil-fuel power plants have significantly lower CO_2 emission rates, as optimists have contended. However, as pessimists have suggested, more efficient plants also emit greater volumes of CO_2, a pattern resembling emission backfires. The fact that backfires occur among the largest sources of carbon pollution (power plants)—and even during a recessionary period when overall demand for electricity waned—is cause for concern. It suggests not only that researchers inquiring into emission rebounds must pay closer attention to the upstream sites of energy production and pollution but also that power plants' rebound effects might grow stronger as the economy recovers.

Equally important, our findings suggest why the EPA should focus on emission levels when establishing pollution targets. Under the Clean Power Plan, states had the right to convert their rate-based goals to level-based goals if they preferred to do so, but the EPA could strengthen targets and also switch to a level-based standard. Under the Affordable Clean Energy Rule, states have to set only rate-based goals, if any. In the past, the EPA has used a level-based standard to assess the environmental consequences of plants modifying their electrical output, setting limits on the volume of nitrogen oxides and sulfur dioxide emitted by plants.[26] The EPA has hesitated to do the same with carbon pollutants in light of court cases (e.g., *United States v. Duke Energy Corp.*, Fourth Circuit, 2005) that have raised questions about whether "modification" in section 111(b) of the Clean Air Act refers to annual emission level or emission rate. However, our results suggest that enhancing plants' thermal efficiency can ironically cause more absolute damage to the climate. Therefore, if the EPA wishes to recommend thermal efficiency as a policy option, it needs to adopt a more appropriate level-based standard to assess its effectiveness.

When we expanded our analysis to include the globe's entire power fleet, our findings are more mixed. Unlike in the United States, results do not support a strong version of the rebound argument that efficiency gains directly increase emission levels. However, they do support a weaker version that says such effects will vary depending on polluters' organizational and global characteristics. Plants that are older, larger, and located in core nations and in nations more tightly connected to the global environmental regime tend to experience rebound effects as plants become more efficient, though these effects rarely outweigh gains made in energy savings, as indicated by the significant direct, negative impact of efficiency on emission levels.

Our international study not only demonstrates that efficiency advocates' policy prescriptions are simplistic but also raises doubts about their understanding of power plants as corporate actors. According to these advocates, power plants pursue the same basic goal of providing a publicly beneficial service (electrical energy), which is achieved by adopting equipment that maximizes efficiency. In contrast, our results suggest that plants also have a vested interest in perpetuating themselves, which may or may not benefit the public. Moreover, plants' ability to adapt and survive is constrained by the forces of inertia. Hence, their performance is determined by internal attributes other than their core technologies, such as their size and age. And because plants do not operate in a vacuum but are embedded in wider material and institutional environments, their behavior is shaped by their location in macroeconomic and cultural systems.

Critics would be correct in pointing out that decisions about how much electricity to produce and with what units are typically made by firms and not plants. To create additional electrical output, for example, a firm's management may choose to switch a certain percentage of the load from its least efficient plant to its most efficient one. As a result, the emissions of the latter plant would increase, but the firm's total emissions would decrease. Our estimated models have begun to address such intrafirm decision-making by nesting plants within firms, but, obviously, more research into this issue is needed. Nonetheless, the fact that increased efficiency is often associated with higher and not lower absolute emissions at the plant level gives us pause.

Overall, our findings in this chapter suggest that for climate policies that emphasize energy efficiency to be fully effective, they must move beyond idealized conceptions of power plants and consider the multiple internal and external characteristics of plants that may condition the effects of technical efficiency on emission outcomes. In the next chapter, we broaden our inquiry to mitigation strategies by testing what the impacts are of several different subnational climate and energy policies and whether their effectiveness varies depending on the mobilization of environmental activists.

5

BOTTOM-UP STRATEGIES

The Effectiveness of Local Policies and Activism

WITH ION BOGDAN VASI

UNDER the Obama administration's Clean Power Plan, power plants were to reduce their carbon pollution by 30 percent from 2005 levels by 2030. The Trump administration scrapped this landmark initiative and replaced it with one that contains no quantitative limits or compliance deadlines at all. It requires only that states consider establishing standards based on a narrow set of operational efficiency tweaks, which our prior chapter revealed could actually increase plants' carbon dioxide (CO_2) discharges. Indeed, the Trump regime favors a new, weaker rule despite its own environmental impact assessment that the planet will warm a catastrophic 7 degrees by the end of this century.[1]

In the continued absence of national legislation on climate change, states have created several policies over the past two decades that, in principle, could curb the electricity sector's climate impact. Some of these policies focus on reducing power plants' CO_2 emissions, while others address this outcome in a more roundabout fashion by encouraging energy efficiency and easing the financial pressures on utilities to maximize their electrical output.[2] However, it remains unclear whether any of these direct and indirect strategies actually mitigate plants' emissions and, therefore, whether states should copy or double down on any of them in the future. Scholars rarely assess the effects that states' policies have on emission outcomes, although on a few occasions they have investigated the effects of these policies on the aggregate CO_2 emissions of states' electricity sectors.[3]

Consequently, researchers stop short of examining whether states' policies reduce carbon emissions at the actual sites where electricity is produced and CO_2 is released—power plants. As our earlier chapters showed, within the electricity sector some facilities pollute vastly more than others, and their emissions may be due to a variety of conditions, including the traits of the plants themselves and the broader economic and cultural environments in which they are embedded. Studies that focus on the effects of policies on state-level outcomes, therefore, overlook variation in power plants' emissions and these potentially confounding factors.

As responsibility for curbing the electricity sector's carbon emissions has devolved to the subnational level in the United States, environmental nongovernmental organizations (ENGOs) have likewise established more local chapters that target specific power plants and lobby their local state officials. In response to complaints that economy-wide approaches to lowering CO_2 emissions are too unwieldy and those focusing on individuals' consumption habits are limited to actors who rarely burn carbon directly, ENGOs have pursed an alternative pathway to a low-carbon future that targets the energy sector, the world's largest and fastest-growing source of emissions.[4]

But just as it has yet to be determined whether states have the capacity to significantly decrease power plants' carbon pollution, there are growing doubts about the ability of ENGOs to reduce the CO_2 emissions of individual plants. Research by world society scholars has shown that from the middle of the last century through the mid-1990s, nations with more memberships in ENGOs tended to have lower CO_2 emissions in the aggregate.[5] However, in the ensuing years, many observers have expressed concern that nongovernmental organizations (NGOs) are increasingly adopting a professional orientation that downplays citizen participation and disruptive politics.[6]

Skeptics note that fossil-fuel industries can exercise considerable power over local citizen groups whose members' livelihoods often depend on the jobs and tax revenues these industries create.[7] Similarly, beginning with Engels's writings[8] on the working poor's exposure to factories' pollution and continuing with the "perpetrator-victim" scenario posited by most environmental justice scholars,[9] research on corporate pollution has tended to conceive organizations as closed systems[10] that are impervious to any pressure from their surrounding communities to improve their environmental performance. It remains to be seen, therefore, whether the local chapters of national ENGOs are sufficiently motivated and resourced today to influence the CO_2 emissions of individual power plants and, if they are, whether their effects are mediated through existing subnational policies or independent of them. And because local ENGOs may be more susceptible to cooptation by the energy industry at the subnational level, it is uncertain whether their local presence helps or hinders the effectiveness of subnational climate policies.

Thus, a number of important questions need to be answered: Which, if any, of the U.S. states' climate and energy strategies reduce power plants' CO_2 emissions? Can civil society mitigate the damage the energy sector is doing to the earth's life-support system at the sites where it is causing the greatest harm—power plants? If so, does civil society directly influence plants' climate-disrupting emissions or impact the situations indirectly through the strengthening of environmental policies? And to what extent might civil society transform policies that have no effect on plants' emissions into ones that do?

Scholars have made little progress in answering these questions for two main reasons. First, as noted in earlier chapters, systematic data on individual power plants and their CO_2 emissions have been lacking. This has forced scholars to rely on crude state-level or national estimates of emissions that obscure differences in pollution caused by individual power plants and make it impossible to ascertain whether plant-level emissions are determined by state policies, local ENGOs, or other factors. Second, researchers have continued to use theoretical frameworks that were tailored for times when climate policies were either nonexistent or largely symbolic in nature. Today, however, due to the growing rationalization of institutional environments, states and markets are held more accountable for climate-disrupting emissions. On the one hand, this creates new opportunities for civil society to pass policies that compel industries to reduce their carbon releases. On the other hand, it is not obvious what role civil society plays in affecting emission outcomes after the policies it advocates have been enacted.[11]

To remedy this situation, in this chapter we use data collected by the U.S. Environmental Protection Agency (EPA) under its Greenhouse Gas Reporting Program (GHGRP) to analyze the effects of U.S. states' climate-focused policies and energy policies with climate implications (explained in the following section) on power plants' CO_2 emissions between 2005 and 2010. Using the same data, we then examine whether local ENGOs can significantly shape plants' emission levels net of effective subnational climate policies *and* change otherwise ineffectual policies into effectual ones. We supplement our quantitative analyses of U.S. power plants with a qualitative one to illustrate the mechanisms through which local ENGOs influence plants' emissions. Prior to conducting our analyses, we briefly discuss the types of policies that states have used thus far to address climate change and sketch a framework for analyzing the direct and mediated effects of local ENGOs on individual power plants' CO_2 emissions.

Contrary to some studies that treat civil society and social movements as occupying distinct spheres, our framework emphasizes how NGOs can possess properties of both. As case studies suggest, characteristic of civil society and the associated world of consultation, NGOs can participate in politics and foster consent through officially approved channels. By the same token, NGOs are sometimes born out of social movements and may provide a space for engaging in forms of dissent not recognized or encouraged by state officials. Rather than relying exclusively on the civil society or the social movement literature, therefore, we seek to provide a richer and more nuanced understanding of NGOs by drawing on both. In the process, we suggest how ENGOs can not only enable certain climate policies to more effectively reduce power plants' carbon emissions but also shape emission outcomes independent of climate policies.

BACKGROUND

State-Level Policies

Although several U.S. states have vowed to forge ahead with their climate strategies despite the lack of federal action and support, the efficacy of their existing policies has yet to be proven. These policies range from emission caps that were primarily motivated to combat emissions to programs like electric decoupling that were created for other reasons but that have implications for plants' climate-disrupting pollution because they try to alter how energy sources are used or managed.[12] For want of a better term, we refer to the latter programs as energy policies with climate implications.

Table 5.1 lists the policies that we tested, the states that adopted them, and the length of time they have been in place (as of 2010). This list approximates the range of measures that states have used to address power plants' carbon emissions. The first set of policies, which are explicitly climate focused, consists of emission caps (including cap-and-trade, a market-based strategy that allows organizations to purchase and trade emissions allowances under an overall limit on those emissions, as well as other CO_2 performance standards that require power plants to purchase technology to limit CO_2 or pay a fine to offset emissions), greenhouse gas (GHG) targets (goals for reducing GHG emissions to a certain level by a certain date), climate action plans (comprehensive strategies for reducing a state's CO_2 emissions), and GHG registry/reporting (systems that require plants to register and record their emissions and emission reductions).

The next set of policies, which are energy related and have implications for the climate, includes efficiency targets, renewable portfolio standards, public benefit funds, and electric decoupling. An efficiency target is a standard used to encourage more efficient generation, transmission, and use of electricity and natural gas. A renewable portfolio standard requires electric utilities to deliver a certain amount of electricity from renewable or alternative energy sources. A public benefit fund provides financial assistance for energy efficiency, renewable energy, and research and development. And electric decoupling is a regulatory strategy designed to ease the pressure on utilities to sell as much energy as possible by eliminating the relationship between revenues and sales volume assumed by market-based approaches. It does so by guaranteeing utilities will receive fair compensation regardless of fluctuations in sales.

TABLE 5.1 STATES' YEARS OF EXPERIENCE WITH CLIMATE-FOCUSED POLICIES AND ENERGY POLICIES WITH CLIMATE IMPLICATIONS, AS OF 2010

Climate-Focused Policies

Emission caps

1–4 years: CT, DE, FL, IL, MD, MA, MT, NJ, OR, WA

≥ 5 years: CA, ME, NH, NY, VT

GHG targets

1–4 years: None

≥ 5 years: AZ, CA, CO, CT, FL, HI, IL, ME, MA, MN, NJ, NM, NY, OR, RI, VT, VA, WA

Climate action plans

1–4 years: AR, CA, CO, FL, IA, KY, MD, MN, MT, NV, NH, NY, NC, OH, PA, SC, VT, VA, WA, WI

≥ 5 years: AZ, CT, IL, ME, MA, MI, NM, OR, RI

GHG registry/reporting

1–4 years: CA, FL, IA, NC, OR, WA

≥ 5 years: CT, DE, ME, MD, MA, NM, NY, RI, VT, WI

Energy Policies with Climate Implications

Efficiency targets

1–4 years: CO, HI, IL, MD, MA, MI, MN, NV, NJ, NM, NC, OH, OR, PA, VT

≥ 5 years: CA, CT, NY, RI, TX, WA

Renewable portfolio standards

1–4 years: AZ, CO, CT, DE, IL, MD, MA, MI, MO, NH, NM, NC, OR

≥ 5 years: CA, HI, IA, ME, MN, MT, NV, NJ, NY, PA, RI, TX , WA, WI

Public benefit funds

1–4 years: AZ, NE, NV, OH, TX, VA

≥ 5 years: CA, CT, DE, HI, IL, ME, MA, MI, MN, MT, NJ, NM, NY, OR, PA, RI, WI

Electric decoupling

1–4 years: CT, ID, MA, MN, NY, RI, VT, WI

≥ 5 years: CA, MD, OR

Note: The renewable portfolio standards examined here are all binding, not voluntary.

Table 5.1 suggests there is considerable variation in the popularity of policies and the length of time states have been using them (through 2010). For example, the climate action plan is the most common policy, having been adopted by twenty-nine states, most of which have one to four years of experience with it. In contrast, electric decoupling is the least common, having been adopted by only eleven states, eight of which have used it for just one to four years.

Scholars disagree over whether these state initiatives are capable of reducing carbon emissions. Some argue that climate-focused policies and energy policies with climate implications can both be effective.[13] Others, including many economists, contend that only climate-focused policies like emission caps can work.[14] And still others claim that both types are too constitutionally constrained or institutionally weak to have a significant impact.[15] Researchers have analyzed the conditions under which states adopt climate-focused policies and energy policies with climate implications[16] as well as these policies' financial costs and benefits,[17] their implications for the development of low-carbon energy technologies,[18] their impact on the share of renewable energy electrification,[19] and their simulated effects on future CO_2 emissions.[20] However, researchers have rarely assessed the effects that states' policies have had on emission outcomes. On the few occasions when they have considered policy effects,[21] they have investigated these effects on the pollution behavior of states and not individual power plants.

Hence, studies have failed to rule out the possibility that the observed effects of states' policies may be explained by features of the plants themselves, such as their size, primary fuel, pollution control equipment, and dispatch systems, and by whether they are publicly or privately owned.[22] In addition to these internal characteristics of plants, there may be external factors or attributes of plants' states and regions that determine the adoption of policies and thus could explain their effects. The association between emission outcomes and states' policies may, for example, be due to the fact that some policies are easier to pass in states where the coal, oil, and gas industries are weak, the Democratic Party exercises more control, the potential for renewable energy is high, energy efficiency is a fiscal priority, cleaner fossil fuels like natural gas have become more affordable, and the regional demand for electricity is growing.[23] Until research determines the net effects of states' policies, it will be difficult for environmental officials to know which of these policies produce effective and generalizable results.

They will also not know whether civil society and social movements in the form of NGOs supplement or condition the effects of policies. At least since Max Weber's foundational writings, sociologists have stressed how civil society can solve problems that the market and state either create or ignore. While acknowledging that the policies promoted by civil society might be adopted but never fully implemented, they suggest that citizen groups can nonetheless be influential within decoupled systems by diffusing cultural models that legitimate social movements, spur corporate action, change government priorities, and reshape people's attitudes.[24] Extending this argument, institutional scholars have contended that in the absence of legislation on the issue of climate change, NGOs can still help reduce the amount of anthropogenic GHGs being emitted by disseminating environmental norms and by pressuring companies and government officials to act.[25] In the next section, we explore this line of reasoning further.

Civil Society, Social Movements, and Carbon Pollution

According to Weber, associational life is the sociocultural basis for political education. As such, it not only is a bulwark for democratic dynamism during modern times when monopolies restrict the expansion of free markets and governments are subject to bureaucratic petrification but also serves to integrate policies and practices in ways that benefit the public at large. Building on these ideas, several scholars have conceived civil society as occupying a middle ground between the state and the market and have suggested how civil society can enlist the state to address threats the market poses to the public. Civil society does this through bonding ties that raise awareness among citizens about their shared interests and through bridging ties that communicate those interests to government, which must adopt policies that reflect citizens' interests to maintain legitimacy.[26]

Still other scholars have observed that market actors sometimes go to great lengths to buffer their core technologies and procedures from state policies by appearing to comply with them, resulting in a delinking of policies and practices.[27] Far from denying the importance of civil society, however, they have insisted that even when the policies it promotes fail to alter business practices, civil society can still bring about real change by forging bridges with market actors and encouraging them to identify with and voluntarily act in the interests of citizens. Applying this logic to the problem of climate change, they suggest that formal environmental policies are not the only mechanism capable of driving emission outcomes.[28] Rather, in the absence of such policies, NGOs can function as "receptor sites" that legitimate and spread environmental norms and discourse. This, in turn, puts informal pressure on domestic actors, including local businesses, to reduce their emissions.

Skeptics contend that businesses may manipulate these same receptor sites to convince civic leaders to adopt a more managerialist perspective that focuses on minor environmental reforms and technical solutions. Especially in communities that depend heavily on businesses that extract, transport, or burn fossil fuels, energy executives may capture local chapters of ENGOs and use them to discourage citizen participation and disruptive tactics. Consistent with the idea that NGOs' autonomy has been compromised is the finding of Doug McAdam and colleagues[29] that NGOs have no bearing on local opposition to pipelines and other large energy projects.

Adding to the uncertainty surrounding the efficacy of NGOs is the growing rationalization and fragmentation of the institutional environment.[30] As rationalization spreads, the disjuncture between policies and practices becomes less tolerated, contributing to fragmentation as government and civic leaders seek to isolate the actors most responsible for undesirable outcomes and develop competing solutions. In federalist countries like the United States, these developments have led to the emergence of a new national culture of accountability accompanied by heightened policy innovation and civic activity at the subnational level. In the area of environmental protection, for example, the EPA now requires industrial plants to make their pollution more transparent by submitting annual reports. At the same time, consistent with this country's tradition of environmental federalism, states and local citizen groups are being encouraged to devise policies that curb industrial emissions and to hold polluters publicly responsible for the threat they pose.

In a period of environmental accountability, the appearance of conforming to ecological norms is no longer sufficient for attaining legitimacy. Governments cannot, for example, rely solely on pollution prevention programs that provide businesses with technical assistance and publicize their participation but that do not assess any penalties regardless of their environmental record. Governments are now expected to develop policies that force major polluters to make real improvements in their environmental performance.

In response to this shift in the institutional environment, businesses will likely still try to insulate their core processes from outside pressures. But instead of relying on previous ceremonial strategies like greenwashing, etc., they will focus more on promoting public policies that (1) use different outcome measures that provide businesses greater flexibility (e.g., emission rates rather than emission levels) or (2) advocate means (e.g., rebates) that seem capable of achieving a particular end (e.g., energy efficiency) even though the end itself may have a dubious relation to the outcome sought by climate activists, which is to reduce or eliminate the absolute pounds of carbon emitted to the atmosphere.[31] An example of the former is President George W. Bush's rejection of the Kyoto Protocol's mandatory cap on GHG emissions in favor of an intensity target. An example of the latter is the EPA's Clean Power Plan, which recommended several strategies to meet the rate-based goals (in pounds of CO_2 per megawatt-hour) it set for states' electricity sectors but did not explain how these goals could be translated into mass-based goals (in pounds of CO_2).[32] In short, during the shift to environmental accountability, it will be easier to recouple policies with practices.[33] At the same time, there may be a growing decoupling of means and ends.[34]

Under these circumstances, environmental activists and ENGOs have an opportunity to forge bridging ties with industry that are more on their terms. Instead of such ties absorbing ENGOs into an institutional logic that privileges industry and quiet incremental change, ENGOs can exploit them to harass management, disrupt routines, generate media attention, and engage in other forms of resistance that have more direct and significant effects on corporate pollution. Because the institutional environment at large expects the state to hold industry accountable for its pollution, ENGOs are also in a better position to implement policies that have real consequences for emission outcomes. But once those policies are in place and they begin to lower emissions, activists themselves may add little to their effectiveness because, again, they are enforced by the larger institutional environment. Similarly, it is unlikely that climate activists and ENGOs will seek to pass or bolster policies favored by industry that use an outcome measure other than a level-based one.

ENGOs may, however, try to bolster policies advocated by industry that promote different ends because, while they may not directly reduce emission levels, in principle they could. That is, ENGOs might realign an end with the latter goal by encouraging businesses to adopt routines that have the best chance of decreasing absolute emissions.[35] For example, as suggested in chapter 4, one of the potential pitfalls of energy efficiency policies is that they enable power plants to economize on their fossil fuels and thus may entice these plants to increase their output to the point where their emission levels actually begin to rise, characteristic of so-called rebound effects.[36] By encouraging downstream users of electricity to keep their consumption levels low, ENGOs can prevent this from occurring and persuade power plants to invest their energy savings in less destructive environmental routines.

ENGOs are likely to have a significant influence on energy companies' decisions because they combine elements of civil society and social movements. As an embodiment of civil society, ENGOs may participate in local and national politics as well as in energy policy making—for example, by joining energy collaboratives and submitting comments to public utilities commissions. At the same time, ENGOs may be involved in the environmental movement's campaigns that pressure electric utilities and elected officials to reduce air pollution and GHG emissions. ENGOs often use social movement tactics—protests, sit-ins, marches, boycotts, etc.—to foster dissent and work outside the institutionalized policy-making system. Therefore, ENGOs should both enable certain climate policies to more effectively reduce power plants' CO_2 emissions and shape emission outcomes independent of climate policies.

While scholars recognize that the environmental movement has had a pervasive influence on the evolution of modern energy systems, it is surprising that the literature on the energy sector "has so often treated [environmental] activists as irrelevant or passive agents."[37] Environmental organizations have clashed with electric utilities and power plant operators since the 1970s. In 1975, for example, the Natural Resources Defense Council (NRDC) sued the Bonneville Power Administration for failing to consider alternatives to constructing new fossil-fuel-powered power plants. NRDC and other environmental organizations also published the *Alternate Scenario* in 1977, which suggested that future electricity demand in the Pacific Northwest could be met primarily with conservation measures.[38]

During the 1980s, environmental groups advocated for demand side management as a solution for the projected increase in electricity consumption. The first collaborative—a plan for energy efficiency and demand side management programs developed in collaboration by environmental groups and electric utilities—was set up in 1988 by an environmental group, the Conservation Law Foundation, and a utility, the Connecticut Light and Power Company.[39] During the 1990s, environmental groups increased their pressure on utilities to go beyond conservation measures and invest in renewable energy as the environmentalist agenda became dominated by the threat of global climate change.

More broadly, research about the influence of social movements on corporations has examined corporate-movement dynamics operating at a national or global level.[40] This research has tended to focus on large multinational companies that have scaled beyond a single community, causing the negative externalities of corporate policies to affect a broad set of geographically dispersed stakeholders. Clearly, as corporations move across state (and national) borders, they become more difficult to formally regulate, leading activists to look for extragovernmental solutions that transcend local communities.

At the same time, much activism continues to be locally oriented, embedded in communities and focused on particular municipalities and local businesses.[41] For example, many manifestations of the civil rights movement involved the organizing of community members to try to change local regulations, rules, and customs.[42] This localism seems especially prevalent among environmental activists, reflecting the specific environmental damages that companies cause to specific geographical areas in the form of toxic dumping,[43] the destruction of lands and resources,[44] and air pollution.[45] We expand this research by developing hypotheses about the effect of local ENGOs on power plants' carbon emissions.

HYPOTHESES

The United States provides an ideal setting for investigating these ideas. The EPA's Clean Power Plan, authorized by section 111(d) of the Clean Air Act (CAA), required the agency to set performance standards for stationary sources of pollution, including power plants. The Trump administration's replacement, the Affordable Clean Energy Rule, is also authorized by the CAA. However, both depart from how air pollutants have been regulated in the past under the CAA in that each seeks to create rate-based as opposed to level-based limits on emissions.

For example, while the Clean Power Plan used 2005 emission levels (measured as total pounds of CO_2) as a baseline for comparison, it also established standards, at least initially, for state-specific emission rates (measured as pounds of CO_2 emitted per unit of electricity produced). Rate-based standards are favored by fossil-fuel industries because they allow for economic growth and thus give power plants more flexibility to improve their environmental performance than do level-based standards. In contrast, ENGOs like the World Resources Institute[46] have stressed the difficulties surrounding the communication and perception of rate-based standards. They note that not only is a rate subject to different interpretations because its denominator can be quantified in several ways (e.g., output versus sales) but also rates tend to decline over time regardless of whether total emissions rise or fall, thus creating a false sense of improvement among the public.

As mentioned earlier, states have been experimenting over the past two decades with policies to lower their power plants' carbon pollution. Some of these policies, which we label direct climate policies, are explicitly climate focused and designed to curb energy-based CO_2 levels by, for example, establishing emission caps or emission targets. Others, which we label indirect climate policies, were created for different reasons, such as to promote the conservation of fossil fuels, but nonetheless may have a bearing on plants' climate-disrupting emissions. Of these, the energy industry least opposes indirect policies because, like rate-based standards, they encourage power plants only to use carbon-intensive fuels more wisely and do not attempt to constrain their productivity.

It follows that to the degree local ENGOs are committed to closing fossil-fuel plants or mitigating the absolute environmental harm they cause, they will put more effort into reducing plants' CO_2 emission levels than their emission rates. Through their public protests and participation in community hearings, local ENGOs can compel plants to reduce their levels or persuade local officials to adopt legislation like emission caps that forces them to do so. But once such direct climate policies are in place, local ENGOs probably add little to their effectiveness. However, through their sharing of technical knowledge, local ENGOs may be able to improve the effectiveness of existing indirect policies by encouraging plants to choose the most environmentally responsible means to conserve and economize on fossil fuels.

Put differently, we can think of local ENGOs as performing two types of roles.[47] As *activist organizations* (exemplified by groups like the World Wide Fund for Nature and Greenpeace), ENGOs use confrontational or outsider strategies (e.g., demonstrations, rallies, lobbying) to confront polluters or advocate for policies that require polluters to reduce their emissions. As *advisory organizations* (exemplified by groups like the Center for International Environmental Law and the Environmental Defense Fund, or EDF), local ENGOs use collaborative or insider strategies (e.g., research-based reports, knowledge construction) to craft policies that are more voluntary in nature and acceptable to industry but that still have the potential to reduce emissions if polluters follow the technical advice these NGOs offer.

Given the paucity of empirical research and theorizing on the efficacy of states' climate-focused and/or energy policies, we are agnostic as to which will exert significant direct effects on power plants' CO_2 emissions. Based on the preceding discussion, though, we do hypothesize the following:

> *Hypothesis 1*: Power plants' CO_2 emission levels are significantly lower in local areas where more local ENGOs are present.
>
> *Hypothesis 2*: The efficacy of states' indirect climate policies, but not their direct climate policies, varies by the presence of local ENGOs.

The appendix to chapter 5 (at the end of the book) provides in-depth descriptions of the data that we analyzed and the methods that we used to test our hypotheses.

ANALYSIS

State-Level Policies

In table 5.2, using ordinary least squares (OLS) regression we see the test results for the effects of the four climate-focused policies on power plants' CO_2 emissions in 2010, controlling for the plants' characteristics, the attributes of the plants' states and regions, and the plants' emission levels in 2005. With respect to plant characteristics, the plants that primarily rely on coal and are large have significantly higher emissions across all models. In contrast, plants founded more recently consistently have lower emissions. Plants that use equipment to control the release of other pollutants have higher emissions. This may suggest that while such technologies curb the emission of other harmful chemicals, they also require more electricity to operate and thus contribute to the discharge of more CO_2.[48] Independent system operators (ISOs) and regional transmission organizations (RTOs), which facilitate more efficient transfers of energy, significantly lower plants' emissions in all four models.

Turning to the other controls, we see that a state's potential for renewable energy and a change in its natural gas prices significantly shape plants' emissions in all four models. Being in a region where electric output is rising (an indication of growing demand for electricity) significantly increases plants' CO_2 emissions in three of the models. As expected, plants' prior emission levels are strongly related to their current ones in every model. (And the F-statistic for the parent group dummy variables is significant, indicating that we can reject the hypothesis that parent companies exert a jointly insignificant effect on emissions.)

Most importantly, we see that net of controls, two of the four climate-focused policies are significant determinants of plants'

TABLE 5.2 REGRESSION ANALYSIS OF THE EFFECTS OF STATES' CLIMATE-FOCUSED POLICIES ON POWER PLANTS' LOGGED CO_2 EMISSION LEVELS, 2010

	Model 1	Model 2	Model 3	Model 4
Plant Characteristics				
Coal fuel (1 = yes)	.294*	.307*	.302*	.319*
	(.141)	(.152)	(.152)	(.151)
Size	.003**	.003**	.003**	.004**
	(.001)	(.001)	(.001)	(.001)
Year founded	–.012***	–.013***	–.013***	–.012***
	(.003)	(.003)	(.003)	(.003)
Equipment to control other pollutants	.290*	.289*	.292*	.283*
	(.148)	(.147)	(.148)	(.143)
ISO/RTO (1 = yes)	–.283*	–.337**	–.334**	–.340**
	(.144)	(.142)	(.143)	(.141)
Public utility (1 = yes)	–.075	–.097	–.133	–.105
	(.393)	(.396)	(.388)	(.384)
State/Region Attributes				
Coal industry influence	–69.308	–66.912	–72.134	–74.784
	(44.326)	(44.764)	(45.113)	(44.195)
Oil and gas industry influence	141.240	138.143	147.521	140.683
	(117.992)	(117.041)	(118.640)	(116.661)
Democratic control (1 = yes)	–.163	–.207	–.193	–.120
	(.193)	(.186)	(.188)	(.193)
% spending on efficiency	–38.195	–68.985	–46.356	–80.167
	(60.629)	(57.959)	(63.197)	(58.386)
Renewable energy potential	–.008**	–.007*	–.006*	–.009**
	(.003)	(.003)	(.003)	(.003)
Change in natural gas prices	.163*	.175*	.213*	.191*
	(.076)	(.086)	(.094)	(.094)
Change in regional electric output	4.809*	4.011*	2.546	3.952*
	(2.350)	(2.192)	(2.839)	(2.008)
Number of other tested policies	–.012	.043	–.095	–.165
	(.092)	(.132)	(.096)	(.087)

Prior Pollution				
Logged emission level in 2005	.824***	.820***	.824***	.818***
	(.041)	(.041)	(.041)	(.041)
Policies				
Emission caps (1–4 years)	–.178			
	(.242)			
Emission caps (≥ 5 years)	–.511*			
	(.233)			
GHG targets (1–4 years)		——		
GHG targets (≥ 5 years)		–.286**		
		(.107)		
Climate action plans (1–4 years)			–.023	
			(.219)	
Climate action plans (≥ 5 years)			–.151	
			(.300)	
GHG registry/reporting (1–4 years)				–.097
				(.267)
GHG registry/reporting (≥ 5 years)				.337
				(.256)
Constant	–22.529	–23.351	–24.038	–22.942
R^2	.776	.774	.772	.772
N	1,129	1,129	1,129	1,129
Number of groups	846	846	846	846
F-statistic of joint significance for group effects	1.88	1.87	1.87	1.86
P-value for F-statistic	0	0	0	0

Notes: Regression coefficients are unstandardized; standard errors are in parentheses; models include group dummies for parent companies.

*p = ≤ .05

**p = ≤ .01

***p = ≤ .001 (two-tailed test)

emissions. Specifically, in states where arguably the most direct measures—emission caps and GHG targets—have been in place for at least five years, plants' emissions are lower. Several of these states participate in the Regional Greenhouse Gas Initiative's cap-and-trade system.[49] In contrast, the two most widely implemented policies examined here—climate action plans and GHG registry/reporting—have no effect. This may be because some climate action plans are just one-off bureaucratic reports and the emission data reported to some GHG registries are not always sufficiently publicized to mobilize local pressure on polluting plants.

In table 5.3, we set out the effects of energy policies with climate implications. Net of the controls, whose effects are essentially unchanged from the results in table 5.2, we see that efficiency targets and renewable portfolio standards have no effect on plants' emissions, whereas public benefit funds and electric decoupling are significant determinants. Efficiency targets may be ineffectual because, if working properly, they decrease demand for electricity as well as for cleaner renewables. Renewable portfolio standards may do little to reduce CO_2 emissions because most renewables are intermittent and, therefore, may still force plants to depend on more reliable carbon-intensive fuels, especially where hydropower or storage technologies are unavailable.[50] That public benefit funds and electric decoupling reduce emissions is consistent, respectively, with the argument that levies assigned to customers' electricity bills can be used to stimulate utility investments in clean energy activities and with the notion that utilities are likelier to engage in such activities when they can make more money by selling less electricity. Importantly, electric decoupling produces more immediate results than the two successful climate-focused policies, providing significant emission reductions in the shorter and longer terms.

TABLE 5.3 REGRESSION ANALYSIS OF THE EFFECTS OF STATES' ENERGY POLICIES WITH CLIMATE IMPLICATIONS ON POWER PLANTS' LOGGED CO_2 EMISSION LEVELS, 2010

	Model 1	Model 2	Model 3	Model 4
Plant Characteristics				
Coal fuel (1 = yes)	.297*	.294*	.296*	.271*
	(.149)	(.145)	(.140)	(.128)
Size	.003**	.003**	.003**	.003**
	(.001)	(.001)	(.001)	(.001)
Year founded	–.012***	–.012***	–.013***	–.011***
	(.003)	(.003)	(.003)	(.003)
Equipment to control other pollutants	.329*	.314*	.330*	.367*
	(.148)	(.148)	(.148)	(.150)
ISO/RTO (1 = yes)	–.236	–.268*	–.192	–.175
	(.149)	(.128)	(.158)	(.156)
Public utility (1 = yes)	–.165	–.124	–.142	–.024
	(.386)	(.387)	(.385)	(.392)
State/Region Attributes				
Coal industry influence	–73.488	–75.689	–84.395	–81.787
	(44.857)	(44.711)	(45.943)	(44.238)
Oil and gas industry influence	84.633	121.065	106.467	128.666
	(126.925)	(125.919)	(123.318)	(123.693)
Democratic control (1 = yes)	–.213	–.254	–.285	.081
	(.190)	(.208)	(.214)	(.250)
% spending on efficiency	–62.311	–77.174	–73.160	–10.640
	(58.916)	(57.950)	(56.198)	(64.924)
Renewable energy potential	–.006*	–.007*	–.008*	–.009**
	(.003)	(.003)	(.003)	(.003)
Change in natural gas prices	.172*	.197*	.251**	.182*
	(.087)	(.087)	(.089)	(.088)
Change in regional electric output	4.602*	5.134	3.757	6.201*
	(2.249)	(2.575)	(2.215)	(2.610)
Number of other tested policies	–.156	–.113	.045	–.093
	(.107)	(.088)	(.104)	(.067)

(*continued*)

TABLE 5.3 (*continued*)

	Model 1	Model 2	Model 3	Model 4
Prior Pollution				
Logged emission level in 2005	.821***	.826***	.831***	.821***
	(.041)	(.041)	(.041)	(.041)
Policies				
Efficiency targets (1–4 years)	.187			
	(.184)			
Efficiency targets (≥ 5 years)	–.029			
	(.272)			
Renewable portfolio standards (1–4 years)		.179		
		(.218)		
Renewable portfolio standards (≥ 5 years)		.018		
		(.251)		
Public benefit funds (1–4 years)			–.367	
			(.238)	
Public benefit funds (≥ 5 years)			–.447*	
			(.224)	
Electric decoupling (1–4 years)				–.528*
				(.245)
Electric decoupling (≥ 5 years)				–.913**
				(.351)
Constant	–21.989	–23.119	–23.601	–19.963
R^2	.774	.773	.771	.774
N	1,129	1,129	1,129	1,129
Number of groups	846	846	846	846
F-statistic of joint significance for group effects	1.87	1.88	1.88	1.86
P-value for F-statistic	0	0	0	0

Notes: Regression coefficients are unstandardized; standard errors are in parentheses; models include group dummies for parent companies.

**p* = ≤ .05

***p* = ≤ .01

****p* = ≤ .001 (two-tailed test)

Environmental Nongovernmental Organizations

Next, we explore the emission consequences of ENGOs. Here we simplified the states' array of policies by subjecting them to a factor analysis to discern which tend to cluster together. Our results, shown in table 5.4, indicate there are two coherent and distinct sets of climate policies. The first set captures whether states have renewable portfolio standards, energy efficiency resource targets, and public benefit funds. We label this factor *indirect climate policies*. The second set captures whether states have emission caps, GHG targets, and climate action plans. We label this second factor *direct climate policies*. In the analyses that follow, we consider

TABLE 5.4 FACTOR ANALYSIS OF STATES' ENERGY-RELATED CLIMATE CHANGE POLICIES WITH VARIMAX ROTATION (N = 50)

	Model 1	Model 2
Emission caps	.343	.624
GHG targets	.334	.641
Climate action plans	.372	.428
GHG registry/reporting	.301	.187
Efficiency targets	.643	.250
Renewable portfolio standards	.807	.285
Public benefit funds	.623	.231
Electric decoupling	.212	.305
Financial incentives for CCS	–.120	.078
Mandatory green pricing	.313	.388
Eigenvalue	2.453	1.465
Alpha (for boldfaced items)	.815	.734

the results of our tests on the two policy packages. (We also use a slightly different model specification in keeping with the recommendations of experts who originally reviewed this analysis.)

Table 5.5 assesses the determinants of power plants' CO_2 emission *rates* over time between 2005 and 2010. Model 1 reveals that plants that rely on carbon-intensive coal as their primary fuel source pollute at increasingly higher rates. This is also true of older plants, which tend to use less efficient technologies. However, larger plants, which reap the benefits of economies of scale, and plants situated in regions where the demand for electricity is growing or in states controlled by Democrats tend to reduce their rates. Net of these controls' effects, the presence of more ENGOs has a negligible impact on emission rates. This is consistent with our argument that environmental activists do not prioritize reducing plants' emission intensities, since that may not decrease the total amount of the carbon they release.

Model 2 adds indicators of states' direct and indirect climate policies. Here we see that only those measures that are specifically designed to address energy-based carbon pollution significantly curb plants' emission rates. That less direct policies are ineffectual is consistent with our findings reported earlier that renewable portfolio standards and energy efficiency programs are insignificant determinants of plants' emissions. In models 3 and 4, the two types of climate policies interact with our measure of ENGOs. In neither case do ENGOs significantly improve the effectiveness of these policy strategies in reducing emission rates.[51]

Table 5.6 uses the same modeling procedure to examine the determinants of power plants' emission *levels*. Importantly, emission rate and emission level for this data set are weakly correlated at .120. This suggests not only that policy makers are wrong in assuming that reductions in rates will automatically result in reductions in levels but also, as we will see, that factors like ENGOs may have different effects on the two emission outcomes.

TABLE 5.5 REGRESSION ANALYSIS OF U.S. POWER PLANTS' CO_2 EMISSION RATES, 2010

	Model 1	Model 2	Model 3	Model 4
Coal fuel (1 = yes)	95.65*	94.24*	93.19*	88.16*
	(52.04)	(52.02)	(52.31)	(52.54)
Size	–63.79**	–63.35**	–63.53**	–63.37**
	(23.57)	(23.61)	(23.67)	(23.63)
Age	5,993.52**	6,111.12**	6,133.19**	6,375.14**
	(2,292.18)	(2,292.21)	(2,298.39)	(2,314.61)
ENGOs	2.32	8.32	21.72	36.31
	(22.47)	(24.94)	(60.55)	(40.24)
2005 CO_2 rate	1,074.81**	1,072.68**	1,072.67**	1,073.35**
	(45.57)	(44.56)	(45.65)	(45.59)
Change in regional electric output	–1,501.80**	–1,774.92**	–1,780.88**	–1,798.91**
	(704.98)	(727.50)	(729.30)	(728.47)
Democratic state	–7.81*	–4.90	–4.31	–4.04
	(7.73)	(7.58)	(7.62)	(7.60)
Fossil-fuel industry influence	–516.51	2,168.56	2,248.50	2,844.96
	(13,703.09)	(14,469.92)	(14,501.13)	(14,500.14)
Change in natural gas prices	31.32	6.71	6.98	10.09
	(28.26)	(33.11)	(33.19)	(33.36)
Population density	–.01	.01	.01	–.01
	(.12)	(.13)	(.13)	(.13)
Median county income	–.04	–.02	–.01	–.01
	(.16)	(.16)	(.17)	(.16)
South central region (1 = yes)	–4.51	–12.37	–10.75	–5.80
	(88.11)	(90.07)	(90.49)	(90.05)
Mountains/plains region (1 = yes)	76.64	95.78	95.60	90.15
	(161.82)	(162.27)	(162.58)	(162.50)
Indirect climate policies		26.48	33.99	26.06
		(33.26)	(45.79)	(33.29)
Direct climate policies		–58.69*	–58.04*	–28.92
		(31.28)	(33.45)	(58.95)
NGOs X indirect policies			–6.31	
			(26.37)	
NGOs X direct policies				–18.38
				(21.70)
Constant	39,555.49	40,555.06	40,704.03	42,503.71
N	1,129	1,129	1,129	1,129
R^2	.73	.74	.74	.74

Note: Standard errors are in parentheses.
*p = ≤ .05
**p ≤ .01 (one-tailed test)

TABLE 5.6 REGRESSION ANALYSIS OF U.S. POWER PLANTS' CO_2 EMISSION LEVELS, 2010

	Model 1	Model 2	Model 3	Model 4
Coal fuel (1 = yes)	.37**	.38**	.37**	.38**
	(.15)	(.16)	(.16)	(.16)
Size	.32**	.32**	.32**	.32**
	(.08)	(.08)	(.08)	(.08)
Age	19.38**	−18.61**	−18.26**	−18.69**
	(6.61)	(6.63)	(6.37)	(6.69)
ENGOs	−.14**	−.12*	.03	−.13
	(.06)	(.06)	(.17)	(.11)
2005 CO_2 level	.77**	.76**	.76**	.76**
	(.04)	(.04)	(.04)	(.04)
Change in regional electric output	6.79**	6.27**	6.20**	6.27**
	(1.96)	(2.03)	(2.02)	(2.02)
Democratic state	−.01	.02	.02	.02
	(.01)	(.02)	(.02)	(.02)
Fossil-fuel industry influence	−24.49	−36.07	−35.28	−36.31
	(38.26)	(40.16)	(40.17)	(40.31)
Change in natural gas prices	.23**	.16*	.16*	.16*
	(.08)	(.09)	(.09)	(.09)
Population density	.01	.01	.01	.01
	(.35)	(.35)	(.36)	(.36)
Median county income	−3.41	−1.89	−9.57	−1.95
	(4.46)	(4.52)	(4.63)	(4.56)
South central region (1 = yes)	−.18	−.13	−.11	−.13
	(.24)	(.25)	(.25)	(.25)
Mountains/plains region (1 = yes)	.50	.52	.52	.52
	(.45)	(.45)	(.45)	(.45)
Indirect climate policies		−.05	.03	−.05
		(.09)	(.13)	(.09)
Direct climate policies		−.25*	−.23*	−.25
		(.13)	(.13)	(.16)
NGOs X indirect policies			−.07*	
			(.04)	
NGOs X direct policies				−.06
				(.60)
Constant	−146.24	−140.56	−138.09	−141.14
N	1,129	1,129	1,129	1,129
R^2	.77	.78	.80	.78

Note: Standard errors are in parentheses.
*p = ≤ .05
**p = ≤ .01 (one-tailed test)

Model 1 shows that the volume of carbon emitted by plants tends to be higher if plants rely primarily on coal, have larger capacities, and are located in areas where the demand for electricity and the price of natural gas are rising.[52] Older plants, which require more repair and, therefore, are operated less frequently, tend to have lower levels of emissions.[53] Importantly, net of these factors, plants are significantly more likely to reduce their emission levels if embedded in counties with numerous ENGOs.

Model 2 reports that ENGOs continue to reduce plants' emission levels after taking into account states' indirect and direct climate policies. The latter set of policies also significantly reduces plants' absolute emissions. These results comport with our first hypothesis that as activist organizations, ENGOs can pressure local plants to reduce their overall emissions independent of whatever climate policies might be in place.

In the next two models, the two sets of climate policies interacted with ENGOs. We found that states' otherwise ineffectual indirect policies are significantly associated with emission reductions if plants are surrounded by more ENGOs. This supports our second hypothesis that as advisory organizations, ENGOs may do little to improve already efficacious direct climate policies, but they can provide the technical expertise that plants may need to voluntarily comply with indirect policies and translate efficiency gains into actual emission reductions.

CASE ILLUSTRATIONS

ENGOs' Influence on Power Plants

To illustrate the mechanisms through which ENGOs may reduce power plants' CO_2 emission levels, secondary data analyses were carried out, using the LexisNexis Academic database to search newspaper articles for various tactics (protest, demonstration, boycott, and lawsuit) used by ENGOs that interacted with electric utilities.[54]

Our search revealed numerous cases in which activist organizations' use of confrontational tactics against a power plant led to either the closure or the retrofitting of the plant. One such example is the case of the Salem Harbor power plant in Massachusetts, which began operating in 1952. The Conservation Law Foundation, a regional environmental organization, mounted a two-pronged legal assault on Salem Harbor Station.

First, it filed a federal lawsuit against plant owner Dominion Energy for repeated violations of the CAA. The lawsuit cited 317 violations of smokestack emission limits between 2004 and 2009 and asked the court to fine Dominion $10.7 million.[55] Second, it organized protests at the Federal Energy Regulatory Commission "to end the plant's reliance on ratepayer subsidies stemming from insufficient planning for reliability."[56] Additionally, environmental groups coordinated a public education campaign aimed at raising awareness about the local air pollution emanating from the power plant; in fact, the Salem Harbor power plant was frequently described by activists as one of Massachusetts's "Filthy Five" power plants. Activists also criticized the power plant for contributing to climate change and for burning low-sulfur coal from Colombia.[57] As the result of their long-term campaign against the power plant, environmental activists achieved a victory in 2012 when the U.S. District Court for the District of Massachusetts approved a consent decree that required Dominion to shut down some of its units by 2014.

Other power plants that have reduced their emissions as a result of ENGOs' confrontational tactics include the Valmont plant in Boulder, Colorado; the Fisk plant in Chicago; and the AES Redondo Beach plant in California.

ENGOs' Influence on Indirect Climate Policies

Secondary data analyses were also conducted to illustrate the mechanisms through which ENGOs may influence indirect climate policies. These analyses entailed examining the literature on the environmental movement and searching newspaper articles using the LexisNexis Academic database to locate evidence of environmental groups' influence on the adoption of indirect policies such as renewable portfolio standards, energy efficiency resource targets, and public benefit funds.

Numerous cases were found in which activist organizations' actions shaped indirect policies. Environmental organizations such as the Union of Concerned Scientists were instrumental in developing the policy framework for California's Renewables Portfolio Standard (RPS) during the mid-1990s. The group of scientists lobbied for the adoption of the RPS in California because "renewables currently cost a little more than fossil fuels and, in a deregulated electricity market, could disappear, taking their many benefits with them."[58] ENGOs played an important role not only in the adoption of the RPS in California but also in the efforts to increase its goals. For example, while the initial RPS required electric utilities to produce 20 percent of their electricity from renewables by 2010, the Natural Resources Defense Council (NRDC) and other environmental groups called on California's policy makers to update the goal to 33 percent by 2020.[59]

Environmental groups, such as Greenpeace, the League of Conservation Voters, NRDC, Public Citizen, and the Sierra Club, have played an important role in the adoption of policies on renewable portfolio standards in many other states, including Texas, Minnesota, New York, Washington, and Colorado. Many of these groups have also published studies that examine the costs and benefits of national proposals on renewable portfolio standards and have prepared suggested legislation on renewable electricity standards.

More broadly, environmental organizations have used collaborative strategies to craft policies that are more acceptable to industry. EDF and NRDC, for example, have worked with electric utilities since the 1970s to encourage the use of conservation measures. In 1976, EDF challenged the Pacific Gas and Electric Company's request to build new nuclear power plants. Using computer models of future electricity demand scenarios, EDF demonstrated in front of California's regulatory commission that customers and investors would benefit more from conservation programs and renewable energy resources than from the construction of new plants.[60] After environmental groups advocated "demand side management" programs and collaboratives throughout the 1980s, over two dozen utilities in ten states worked with them by 1991 to reduce energy consumption through these programs.[61] The Midwestern Power Sector Collaborative was also formed as part of an ongoing collaboration between ENGOs and electric utilities.

This brief examination of specific cases allowed us to identify some of the processes through which ENGOs have contributed to a reduction in power plants' CO_2 emission levels. It also allowed us to understand how ENGOs have influenced indirect climate policies. In some cases, ENGOs have used protests and demonstrations to pressure local and state elected officials to act; in other cases, ENGOs have sued electric utilities for violating the CAA; in still other cases, ENGOs have published reports and worked with electric utilities and state agencies to adopt and implement renewable portfolio standards, energy efficiency resource standards, and public benefit funds. While space limitations prevent us from providing a detailed analysis of all the strategies ENGOs use, it is important to mention that activists' actions were often coordinated with local chapters of national ENGOs. For example, anticoal activists from around the country received support from the Sierra Club's Beyond Coal campaign or from Greenpeace's Quit Coal campaign.

DISCUSSION AND CONCLUSION

The findings in this chapter reveal that certain policies devised by U.S. states to mitigate climate change have, in fact, reduced the emissions of the largest sources of GHGs—power plants. Contrary to many energy and environmental economists who insist that market-based policies are the only viable means to reduce plants' CO_2 emissions, we found that electric decoupling has been quite successful. And contrary to skeptics who argue that ENGOs have been coopted by energy executives and have compromised their environmental goals to focus on lowering the rate rather than the level at which plants pollute, we found that even in areas where the fossil-fuel industry is strong, ENGOs tend to suppress the total pounds of CO_2 that plants emit. This suggests that utilities are relatively open systems[62] and that their environmental performance can be influenced by a mobilized local citizenry.

In addition, we found that ENGOs curb carbon pollution independent of policies like emission caps and GHG targets, which explicitly address the carbon emissions of power plants, and that they enhance the effectiveness of others like renewable portfolio standards and energy efficiency programs, which encourage the economical use of fossil fossils. So while we found evidence that the latter, referred to here as indirect climate policies, fails to reduce plants' CO_2 emissions, this also depends on the civil context in which plants are embedded.

Finally, our qualitative analyses identified several tactics used by local ENGOs that have directly or indirectly reduced plants' carbon pollution, including outsider strategies like organized protests, lawsuits, and petition drives as well as insider approaches that involve, for example, the sharing of technical knowledge. Through these mechanisms, civil society is reducing emission levels (as opposed to rates) and motivating plants to adopt real green practices.

In addition to providing new insights into the ways state policies directly and indirectly shape emission outcomes, our study in this chapter makes several important contributions to the literatures on global normative systems, social movements, and environmental sociology as well as public policy. We advance research on global normative systems by shifting attention from aggregated environmental outcomes to the sites where the disjuncture between policies and practices is most likely to occur—power plants. By combining information on local ties to NGOs with data on plant-level outcomes, we have uncovered some of the conditions under which individual plants are more or less likely to comply with or decouple from the expectations of such environmental institutions.

We extend social movement research that has focused on the diffusion of environmental policies by investigating whether policies and ENGO activists actually deter pollution. And in keeping with ecological modernization theory within environmental sociology,[63] our research speaks to how environmental organizations can tap into the growing ecological concerns of developed countries to bring about real change in corporate environmental behavior. Finally, our research in this chapter suggests that state policy makers must not simply solicit the input of citizen activists in devising climate policies but also encourage the long-term involvement of activists to ensure that policies are effectively implemented.

Of course, ours is not the final word on the efficacy of particular policies. Some of those we found to be ineffective as of 2010 might eventually become significant determinants. For instance, the first compliance period for several policies on renewable portfolio standards had not occurred by 2010. Conversely, policies adopted before the 2005 base year may have reduced emissions in the years leading up to that point. We do note that studies conducted after 2010 largely comport with our results here.[64]

In addition, our study does not try to determine the optimal way to design individual policies. More research is needed on whether certain variations on a policy—for example, whether renewable portfolio standards allow the trading of renewable energy credits and/or have aggressive, binding targets[65]—make a difference in power plants' environmental performance. Likewise, more studies are needed to determine the most effective policy combinations[66] and the mechanisms through which policies affect emissions. For instance, energy efficiency targets may reduce emissions when bundled with electric decoupling. With respect to mechanisms, renewable portfolio standards may reduce emissions by moving electricity production away from fossil-fuel plants or by raising the cost of electricity.

We also recognize the importance of overall carbon outputs; increased efficiency at the plant level could be negated if the total number of power plants or the amount of electrical output increases in the future. And our study does not address potential problems involving "carbon leakage" or interactions with other climate-related policies like fuel economy standards.[67] In addition, although our models control for state-level changes in natural gas prices and plants' dispatch systems (ISO/RTOs), the effects of lower gas prices may be more complex than what our models capture. Other studies estimate that between 2005 and 2010, the U.S. electricity sector's CO_2 emissions dropped by

6 percent while its carbon intensity fell by 2.5 percent,[68] suggesting that much of the decline in emission levels was due to electricity mix switching. Future studies will, therefore, need to tease out the direct effects of fuel switching on plants' emissions from the indirect ways that states' policies facilitate the shift to natural gas. Finally, additional research is needed to understand which social movement tactics (e.g., protests, petitions, sit-ins, lobbying) are used most often and which ones are most effective in reducing GHG emissions and other environmental harms.

6

NEXT STEPS

Future Research and Action on Society's Super Polluters

In late September of 2019, one of the authors of this book and his youngest daughter joined in the Global Climate Strike. The largest protest of its kind, this worldwide event drew more than 4 million individuals from over one hundred countries, all demanding that action be taken to arrest climate change. In Denver, where the author and his daughter participated, an estimated seventy-five hundred people marched from the city's Union Station through the 16th Street Mall to the Colorado State Capitol Building. As the protesters poured into the street, their chants of "Save our planet" and "The time is now" echoed off high-rise buildings. Office workers in natty attire left their desks to observe the protest through glass walls above. A few shoppers paused to take selfies with marchers in the background. And a maintenance crew laboring over a pothole looked up and smirked.

Several of the youngest demonstrators, who had walked out of their classrooms that day to attend, were interviewed by reporters along the way. They expressed dismay at people standing idly by while the planet burned and described their fears about the future. As one youth activist put it, "People keep saying we need to preserve the Earth for our children and their children and their grandchildren, and I'm thinking, is the Earth going to live that long? I'm semi-prepared for a future that's post-apocalyptic. I know that's sad, but it's true."[1]

Aside from such heartfelt concerns and his deep appreciation for his daughter inviting him to the event, two things about the event especially stood out in the author's mind: the signs carried by the marchers and the way the march concluded. The signs all alluded to the precarious state of the planet and its inhabitants but attributed this condition to very different things. They blamed factors ranging from economic systems ("For the Planet to Live, Capitalism Must End"), selfishness ("We See Your Greed"), and ignorance ("Face Facts") to politicians ("Don't Trump Our World"), generations ("You'll Die of Old Age, We'll Die of Climate Change"), and a lack of connection with the planet ("Love Your Mother"). The closest any of the signs came to mentioning an actual large-scale emitter of greenhouse gases (GHGs) was when they alluded to the suppliers of carbon inputs or the fossil-fuel industry ("Keep It in the Ground").

The other thing that stood out was that after the protesters arrived on the lawn of the state capitol building, many were unsure what to do next. Although a podium had been set up at the steps of the capitol building where speeches were being given, very few gathered around it. Instead, marchers mulled about, checking their smartphones and asking whether they should leave and go back home, which several did. It was an awkward and anticlimactic finish to a historic protest. There is no doubt other marches that day finished on a higher note. And perhaps if organizers had distributed instructions beforehand about what to do at the march's conclusion or if Greta Thunberg, youth climate activist and *Time*'s Person of the Year, had been one of the speakers, more would have congregated at the podium.

Still, the fact the march fizzled out at the end was striking and consistent with a point made by social movements scholars[2] that unless mobilized citizens can agree on a villain, sustaining a movement and achieving its goals will be especially challenging.[3] Reading scientific reports about the effects of a problem or experiencing those effects firsthand might be enough to enrage citizens and motivate them to take to the streets. But if that rage is not aimed at a specific actor or group of actors directly responsible for the harm done, mobilized efforts can flounder.

Similarly, we have argued that scholars and government officials have been slow to acknowledge the carbon elephants in the room. The utility sector emits the most heat-trapping carbon dioxide (CO_2). And yet climate researchers and policy makers have failed to address the subset of power plants disproportionately accountable for those releases. As a result, their assessments too often lack the empirical realism needed to tackle the problem of climate-disrupting emissions.

To rectify this shortcoming and shift attention to the worst of the worst, we have outlined a super polluters framework that illuminates the sites, causes, and mitigation of electricity-based climate change. This approach contrasts with the conventional one espoused by energy and environmental economists, which dismisses differences in power plants' environmental harm as an artifact of differences in their productivity, attributes high-carbon electrical generation to price factors, assumes improvements in technical efficiency will reduce both the rate and the level at which plants pollute, and advocates market-based mitigation policies free of outside intervention.

Consistent with our framework's predictions, which we laid out in chapter 1, we found that not only are emissions distributed unevenly across power plants in nations throughout the world, but also these disproportionalities cannot be explained by variation in plants' electrical outputs. Plants pollute at extremely high levels and rates partly due to a host of interconnected social structures. Depending on their internal and external properties, more efficient plants can actually have higher carbon emission levels. And some non-market-based policies effectively reduce power plants' carbon pollution, especially where climate activists have created a strong local movement infrastructure.

FUTURE RESEARCH

These findings raise a series of important questions for future research. First, given the empirical reality of polluter disproportionalities, *how can hyperemitters be incorporated into the perspectives that currently shape the thinking of the Intergovernmental Panel on Climate Change (IPCC) and other bodies that assess climate change science?* These organizations are instrumental in drawing attention to disparities in carbon pollution across nations and sectors. But they rarely probe such differences at the level where energy is physically produced and carbon is released, mainly because the intellectual paradigms they draw on do not consider or promote such inquiries.

As noted in prior chapters, according to the conventional wisdom of energy and environmental economists, extreme emitters can be disregarded because they are statistical outliers and presumably deliver more electrical services. Other formulations of the anthropogenic drivers of carbon pollution, which are increasingly popular with groups like the IPCC, might at first glance seem capable of providing a corrective and reining in super polluters. Agent-based modeling, for instance, seeks to go beyond purely economic explanations and complicate strong assumptions about rational actors by offering more fine-grained analyses that differentiate actors and their motives. However, like the conventional approach it hopes to supplant, it assumes that environmental well-being is proportional to economic well-being.[4] Hence, it, too, fails to integrate messy real-world actors like profligate polluters into its elegant, stylized emission and mitigation scenarios.

In general, the perspectives that have most influenced entities like the IPCC focus on environmental problems as opposed to environmental privileges, and their shared tenet that extreme polluters are not problematic has been repeated so often, with so little challenge, that it is now taken for granted and forces questions about disproportionalities to essentially disappear. Considering how these frameworks and their modeling techniques greatly shape the public's understanding of climate change, incorporating super polluters seems prudent and would enable policy makers to make substantial environmental progress by selectively regulating these actors.

Second, with respect to the synergistic effects of social structures, *what historical processes explain why certain super polluters have certain structural profiles? What processes are shaping the profiles of future ones? And what are the profiles of their green counterparts and the developments that facilitated their creation?* Such questions are consistent with what Monica Prasad[5] has recently termed a "problem-solving sociology," which uses comparative methods to identify the causes of wicked problems and not just their consequences (i.e., the villains and not just their victims), as it is only through uncovering their causes that we can identify effective solutions.

Agent-based modeling, IPAT,[6] POETIC,[7] and similar perspectives shine a light on the social determinants of carbon pollution by examining the independent effects of macrolevel factors like population, affluence, and urbanization on aggregate emission outcomes. Our findings go several steps further in demonstrating how global, political, and organizational factors combine in complex ways to shape the environmental behavior of individual polluters. Still, we do not know what decisions and circumstances contribute to these configurations. Nor do we know what forces are shaping the makeup of tomorrow's super polluters. Not only were industry and government likely guilty of facilitating the creation of super polluters, but also community leaders, neighbors, and even environmentalists were probably unintentionally complicit. In his rich analysis of solid waste management in Chicago, David Pellow[8] describes how many of the businesses that handled waste (and endangered poor and minority neighborhoods) were founded and staffed by first-generation immigrants and African Americans, illustrating the complicated relationship between dirty polluters and the communities most affected by them. As new coal plants are being erected in rapidly developing nations such as China and India, researchers need to identify the parties enabling new super polluters.

Conversely, scholars also need to investigate the heroes of the electricity sector (and others) that have successfully lowered their emissions.[9] For example, the Nanticoke coal-fired plant in Ontario, Canada, which was the largest coal plant in North America from the early 1970s through the early 2010s, reduced its carbon emissions more than 40 percent from 2004 to 2009. The sharp reduction in emissions followed nearly $900 million in investments to make the plant more energy efficient. Over the next five years, the energy produced by Nanticoke dropped nearly 50 percent while the heat rate fell by 20 percent, suggesting a less active but also more efficient plant. In addition, the province announced a plan to phase out all coal plants by 2009 (though this plan was abandoned in 2006). The Nanticoke plant was fully decommissioned in 2013 and, in its stead, sits a 44-megawatt (MW) solar farm built in partnership by Ontario

Power Generation, the Six Nations of the Grand River Development Corporation, and the Mississaugas of the Credit First Nation.[10]

In short, a combination of regulatory control, civic activism, and financial investment created the conditions necessary to decommission a super polluter and replace it with renewable energy, but it would be a mistake to extrapolate from one case alone. The Nanticoke power plant is one among more than ninety-seven hundred of the world's power plants that reduced their emissions between 2004 and 2009, the majority of which likely did so by decommissioning individual units and (if burning coal) by retrofitting or replacing them with natural gas–burning units. Which combinations of energy policies, consumer demands, activist pressures, and fuel prices produce heroes is unknown.

Third, regarding the paradoxical impact of efficiency, *how can super polluters produce more output with fewer inputs but still decrease their emissions?* Past research on emission rebounds suggests the answer is quite simple: to ensure that efficiency-enhancing technological improvements reduce fuel use, efficiency gains must be paired with green taxes, cap and trade, higher emissions standards, and other forms of government intervention that reduce demand. And yet our analyses reveal that after taking into account demand factors like fuel costs and electricity prices, backfires can still occur, especially among older and larger plants that are locked into established routines and plants that are located in powerful national economies that pay lip service to environmental ideals.

This suggests it is not enough simply to tax away the cost savings from efficiency gains or otherwise remove these savings from further economic circulation. They may need to be captured for reinvestment in natural capital rehabilitation, as Matthis Wackernagel and William Rees[11] suggest, to have their intended environmental impact. Or policy makers should abandon one-size-fits-all energy efficiency mandates entirely and instead couple required efficiency improvements with nonmarket interventions tailored to offset plants' harmful entrenched practices and social circumstances. Until interventions are socialized in some such way, energy efficiency initiatives are unlikely to produce the win-win outcomes the climate movement hopes for and will legitimate the continuation of environmentally destructive behavior.

Fourth, in terms of local policies and citizen engagement, *which social conditions weaken otherwise effective subnational energy and climate policies?* While discovering that the local chapters of environmental nongovernmental organizations (ENGOs) enhance the efficacy of policies is encouraging, it may also be true that some actors and circumstances contravene them. For example, in a historical analysis of policy formation in the U.S. electrical energy industry, Harland Prechel found that political mobilization of powerful organizations with economic agendas tends to cripple existing policy (and presumably to worsen environmental quality).[12] And after state structures are put in place to enforce environmental policy, corporations are often able to coopt them to provide sociopolitical legitimacy for their economic interests.

As nations like the United States retreat from their climate pledges and are captured by companies that produce or burn fossil fuels, cities, regions, and other subnational actors have taken the lead in developing policies to reduce anthropogenic GHG emissions.[13] They seek to generate public pressure on nations that so far have resisted stringent measures and thus to ratchet up commitments. The success of their efforts is crucial because the Paris Agreement is organized as a bottom-up structure that allows individual nations and subnational actors (e.g., states, counties, municipalities) to voluntarily set their own mitigation targets and choose the measures to achieve them. This polycentric policy regime is also symptomatic of an emerging experimentalist governance system[14] in which local political actors set goals for themselves and continually revise them based on comparisons with the performance of alternative approaches in other contexts. While this decentralized system seems well suited for tackling wicked problems like anthropogenic climate change during a time when power is rapidly diffusing in the international system, it has potential drawbacks. As Jamie Peck and Nic Theodore[15] argue in their writings on "fast policy," policy decisions that are found to work in one jurisdiction and promoted as "silver-bullet solutions" may be quickly copied by others without an objective analysis of their efficacy in dissimilar settings.

Hence, the effectiveness of subnational policies can be severely compromised by local circumstances and problems. For example, as related research on environmental justice, innovation, and corporate power suggests, energy and climate policies may be less likely to decrease power plants' emissions in areas with more inequality, less entrepreneurial activity, and domestic electricity markets dominated by a small number of utility companies.[16] As the literature on hybrid logics also suggests,[17] climate policies that impose a high price on carbon emissions and energy policies that indirectly address emissions by promoting energy efficiency or the development of alternative energy sources may work less well together within moderately centralized governance systems.

With funding from the National Science Foundation we are currently investigating these and other issues. Our new project promises to shed new light on the efficacy of subnational mitigation programs during a period when energy usage and carbon pollution threaten to grow in magnitude. We also hope to demonstrate that despite the seemingly obvious alignment between fast policies and a rapid social world, there are clear policy benefits to be had from employing a more contextualized and evidence-based "slow policy" strategy.[18]

OTHER SUPER POLLUTERS AND MITIGATION OPTIONS

Having suggested that substantial environmental gains might be achieved by focusing research and mitigation efforts on hyper-emitting power plants, we are quick to acknowledge there are other actors that have been labeled super polluters that could be targeted. For instance, some suggest that the most effective way to address climate change is to take aim at "carbon majors," the one hundred companies responsible for supplying 70 percent of the world's fossil fuels.[19] Still others contend a better strategy would be to go after the "polluter elite," individuals who own large shares of carbon majors and/or are among the richest 10 percent of people in the world.[20] According to today's most famous social scientist, Thomas Piketty, the latter group is responsible for 45 percent of global emissions.[21]

Researchers and activists have also proposed strategies for combating these super polluters. Groups targeting carbon majors recommend filing climate liability lawsuits against investor-owned companies like BP, Chevron, ConocoPhillips, and Exxon for misleading shareholders about the impact of climate change on company profits as well as against governments for denying citizens' rights to a clean and safe environment.[22] Those targeting the polluter elite advocate imposing a global progressive carbon tax on the top 10 percent of world emitters (i.e., individuals emitting more than 2.3 times average world emissions) or on certain items associated with high energy consumption such as cars, air-conditioning systems, and air transport.[23]

Unfortunately, these tactics have had limited success so far. Because the world's richest possess vast resources to resist measures that threaten their privileged status and many aspire to join their ranks, attempts to tax their lifestyles have gained little traction. Although an increasing number of plaintiffs and lawyers have tried to use the courts to punish fossil-fuel suppliers, there has been considerable pushback from trade groups like the National Association of Manufacturers that claim they are suing to "stuff their already-fat pockets."[24] They and others suggest that the proper path for lowering carbon emissions includes innovation and collaboration—not litigation. Federal judges who have been sympathetic to these arguments have dismissed climate liability lawsuits by New York City, San Francisco, and Oakland on the grounds that the U.S. Environmental Protection Agency and not state courts is responsible for regulating GHGs. These cities are appealing the dismissals, but this situation illustrates the uphill battle that climate advocates face in the legal system.

At the same time that questions are being raised about the feasibility and legality of reining in profligate consumers and fossil-fuel producers, other developments and factors point to the possibility of targeting the super polluters studied here—power plants. Organizations like the Energy and Policy Institute[25] have uncovered evidence showing that as far back as the late 1960s, utility companies understood how burning fossil fuels impacts climate change. Nonetheless, industry associations such as the Edison Electric Institute and Electric Power Research Institute still pushed coal, the largest emitter of CO_2 among fossil fuels, as an energy solution. This information is similar to what has been uncovered about the oil industry: that it was warned about the potentially catastrophic effects burning carbon would have on the climate but worked to create doubt about that science and opposed government action to combat climate change.

The electric utility industry has funded special interest groups, such as the U.S. Chamber of Commerce, that attack climate science concerns as based on "religion" and not "scientific facts." This continues despite growing support among citizens[26] and even investors for limiting CO_2 from power plants. In 2017, for instance, investors in major utilities, including AES, Dominion Energy, DTE Energy, Duke Energy, PPL, and Southern Company, called for them to develop plans to align their business models with the Paris Agreement goal of limiting global warming to 2°C or less; however, the utilities refused.

The fact that utilities, unlike carbon majors and the polluter elite, directly burn and emit carbon may make their culpability easier to establish.[27] New York State's Office of the Attorney General, for example, reached settlements with three major electric utilities after investigations revealed they had not adequately disclosed to investors the financial risks posed by their power plants' emissions. Still, the ability to sue power plants will likely hinge on whether courts determine the Clean Air Act and similar legislation can preempt state tort liability for climate change-related damages. The Urgenda Foundation recently won a series of court cases in the Netherlands that required its national government to reduce that country's CO_2 emissions by at least 25 percent by the end of 2020 compared to 1990 levels.[28] Whether courts in other nations will rule similarly, thus making it easier to sue individual plants, remains to be seen.[29]

Given the uncertainty surrounding the effectiveness of tactics like liability lawsuits, climate advocates will need to consider alternatives when asking *What would it mean to target hyperemitting power plants?* Knowing who these villains are and where they are located is crucial. At the same time, the more we know about them, the greater the challenge of reining in their emissions might seem. Another of the authors of this book lives less than an hour from two of the dirtiest power plants in the United States—the Scherer Steam Generating Station in Juliette, Georgia, and Plant Bowen in Cartersville, Georgia. In 2009, the Scherer and Bowen plants ranked first and third, respectively, in the United States in their total carbon emissions. (The second-dirtiest plant, the James H. Miller Jr. Electric Generating Plant, is less than three hours west in West Jefferson, Alabama.) It is easy to get discouraged about the dangerous consequences of a warming climate when living in such close proximity to three of the world's most egregious super polluters. Yet, in 2019, Georgia Power updated its twenty-year energy plan and called for the closing of five coal-burning units in the state, including two at Plant Bowen. To be sure, much of the capacity lost through closing coal units will be replaced by natural gas–burning units, but Georgia Power has also increased its solar holdings, with some estimates suggesting solar could account for nearly 20 percent of the company's peak summer energy capacity in the next five years.[30]

So what should be done about these types of facilities? How can hyperemitting power plants' emissions be leveled or, perhaps more crucial, reversed altogether? From the vantage point of our framework and analyses, there are at least five options available to begin this process. First, activists and civil society groups could organize more protests aimed at particular plants and their parent companies. Results suggest this strategy may be especially effective where energy policies designed to indirectly curb plants' emissions, such as renewable portfolio standards and efficiency targets, are already in place. Focusing on the dirtiest power plants that are subject to these policies could enable activists to attribute blame and mobilize anger, which Doug McAdam[31] and other social movement scholars argue are critical to sustaining collective action as well as to enhancing movement participants' sense of efficacy. Participants in the divestment movement, such as 350.org, are beginning to embrace this tactic and target super polluters like the Taichung power plant. This strategy can trigger primary stakeholder activism against a plant and/or its owner by shaping how a facility's environmental risk is perceived by investors.[32]

Second, instead of just focusing on the plants that emit the most carbon pollution, activists could devise policies based on the attributes that make them dangerous in the first place. Our fuzzy-set qualitative comparative analysis (fsQCA) of the social determinants of power plants' emissions revealed that four distinct combinations of global, political, and organizational factors not only distinguish plants with the highest emission rates but also predict those with the highest emission levels. According to fsQCA practitioners,[33] remote conditions are usually less amenable to change because they are, by definition, temporally or spatially removed from the outcome being explained. On the other hand, immediate conditions are normally more amenable because, theoretically, they have closer connections to the outcome of interest. When deciding how to reduce a particular super polluter's emissions, therefore, it would be wise to focus on the immediate elements that fsQCA suggests must be in place for facilities of that type to pose a risk. Consider, for example, the second type of super polluter identified by fsQCA—quiescent plants, or those owned by dominant utilities in normatively disengaged core countries. It is probably logistically and legally impossible to physically remove a plant from a nation that has few ENGOs and is centrally located in the global economy. However, it may be possible to create incentives to diversify ownership in an electricity market so that utilities are more responsive to local concerns about climate change.

Third, societies could increase their efforts to implement the policies that our empirical results indicate have a direct mitigating effect on power plants' emissions. Again, measures like emission caps, GHG targets, public benefit funds, and electric decoupling were found to exert potent independent impacts. One potential problem with this strategy is that, because it is based on findings from a sample of U.S. plants and models that estimate how emissions deviate from those of the statistically average plant, it may be less applicable to non-U.S. plants and those at the tail end of the emissions distribution. Nonetheless, this strategy might offer a useful starting point for designing policies whose ultimate aim is to stop super polluters.

Fourth, wealthier nations or international entities like the United Nations could establish something akin to a Green Marshall Plan that would provide funding and financing to cover the costs of refurbishing, retrofitting, or replacing the world's worst super polluters. Many utilities no doubt would resist such an effort, citing the fact that they invested in their generators in good faith, regulators approved these investments, and shareholders expect rewards for taking on significant investment risks. However, an increasing number of utilities are agreeing to retire their plants early in exchange for compensation of some kind, suggesting they may be amenable to similar measures. In Colorado, for example, pending legislation would offer ratepayer-backed bonds to securitize stranded coal plants and generate savings for towns where they are located.[34] One challenge to implementing this strategy, however, is estimating the possible costs involved, which utilities often refuse to disclose publicly. However, tools such as life-cycle analysis have been used to generate decommissioning estimates and may provide a useful way to circumvent this problem.

Fifth, replacing or retrofitting super polluter power plants could be the centerpiece of major infrastructure projects. Given that very few of the individuals who still deny the reality of climate change will likely be persuaded by additional scientific evidence, one way past the political impasse on climate change might be to create a "large climate-industrial complex"; for example, national governments could take over their electric grids, a move that would offer immediate benefits to large numbers of voters analogous to those provided by President Franklin Roosevelt's New Deal. Alternatively, instead of nationalizing entire electricity sectors, as some have proposed, it may be more politically feasible initially for governments to take ownership of only the most egregious polluters. Converting this small subset of facilities into democratic utilities could go a long way toward decarbonizing a nation's economy, especially if other sectors that still rely on combustion-driven technologies, like gasoline vehicles and natural gas heating and cooling systems, were to simultaneously switch to electricity-powered alternatives, like electric vehicles and heat pumps.

Finally, information about hyperpolluting plants could be used to develop carbon tax schemes and supporting materials that would improve the likelihood of their passage. Although insufficient by itself, raising the cost of burning fossil fuels is considered the most efficient way to mitigate climate change. However, in the wake of France's "yellow vest" protests against sudden hikes in the price of petrol and Washington governor Jay Inslee's failed attempts to put a price on carbon, many have concluded such a policy is not politically feasible.[35] Still, the issue is likely to be a topic of national debate in the near future as the oil and gas industry, sensing a U.S. climate policy is inevitable and wishing to secure the gentlest and most predictable energy transition possible, has recently supported a carbon tax bill introduced in the U.S. House of Representatives. This bill would impose a carbon tax starting at $40 per ton of carbon emitted[36] and "rising gradually" at an as-yet-unspecified rate, with all the revenue returned as dividends or rebates to individual citizens.

Environmentalists consider several features of the bill unacceptable, including requirements that the tax be a replacement for all other regulations and that dividends go only to taxpayers (as opposed to other mitigation efforts). But even if environmentalists can get the bill's proponents to compromise on these issues, there are other aspects of this bill and others like it that, if left unaddressed, will limit their effectiveness and chances of approval.

First, although a carbon tax should ultimately be applied to a nation's entire economy, doing so from the outset could prove to be unwieldy, as some sectors are better positioned than others to reduce their emissions. Second, the point along the energy supply chain at which a carbon tax is levied is not specified. Typically, such taxes are assessed downstream, where fossil-fuel-based energy is consumed, or upstream, where fossil fuels are extracted. However, the former requires a very complicated bureaucratic system and would still not capture imported fossil fuels, whereas the effects of the latter can spill over into other industries that use coal, gas, and oil. Third, carbon tax bills supported by the oil and gas industry seem unlikely to extend the fight against fossil fuels beyond the elimination of coal. This is an important consideration because, while cheap natural gas is reducing emissions by displacing coal in the United States and Europe, natural gas was the planet's fastest-growing fossil fuel in 2019, and emissions from it are rising much faster than carbon emissions overall.[37] Fourth, not only is the rate of the tax increase left unspecified, but also the same per-pound tax is applied to all polluters. While such a flat tax might seem equitable, it ignores how economies of scale enable larger emitters to externalize pollution at lower costs and how substantial progress toward a zero-carbon economy could be achieved by simply assigning a heavier tax to them.

We argue that a carbon tax would be most effective and least disruptive if it was initially limited to the sector where the most cost-effective reductions can be found—the electric power industry. A carbon tax on just this sector would achieve 90 percent of the reductions of an economy-wide tax in 2020.[38] This is because electricity's price elasticity is greater—i.e., a relatively low tax will lead to more fuel switching and changed behavior in this sector than in others.

A carbon tax on electricity would also be optimized if assessed "midstream," or at the power plant. In several countries, plants already collect and report environmental data, so adding GHG pollution should be a relatively low administrative burden. Plus it would be more manageable because there are vastly fewer fossil-fueled power plants in the world than there are oil wells, fracking sites, and coal mines—and especially individual consumers. Taxing power plants also directly sends a strong price signal, which should encourage them to make investments in renewable energy technology.

In addition, our analysis suggests that focusing on the power plants that emit disproportionate amounts of carbon as opposed to seeking to keep one type of fossil fuel (coal) in the ground would ensure a continuing reduction in the release of CO_2, particularly at the high end of the pollution distribution. And, finally, we contend a progressive carbon tax with tiered rates that charges the biggest polluters more for each ton of carbon they release would not only rapidly and extensively decarbonize the electricity sector but also be more politically feasible. The vast majority of plants, which pollute at relatively low levels, would be less burdened by it than a flat rate tax.[39]

It is difficult to know which, if any, of these options society will choose to pursue, especially considering the ongoing and likely changes occurring within the realm of energy generation. For instance, the IPCC recently reported that renewables would need to supply between 70 and 85 percent of global energy by midcentury to prevent a 1.5°C temperature rise,[40] and coal-fired capacity elsewhere has, for the most part, risen in the past decade. According to the Global Coal Plant Tracker, global coal-fired capacity grew from 1,603 gigawatts (GW) in 2010 to 2,024 GW in 2018, although the flow of proposed and constructed coal plants in the pipeline has begun to slow.[41] By far, the largest share of new coal-fired capacity is in China, which increased capacity from 197,431 MW in 2001 to 972,514 MW in 2018, and, to a lesser extent, in India, which increased capacity from 61,564 MW in 2001 to 220,670 MW in 2018. Plant retirements continue to rise in the United States and throughout Europe, and according to the International Energy Agency,[42] global coal investments have stalled and are now declining given the saturation of the market in China.

Despite these uncertainties and developments, we are encouraged by the fact that we are not alone in casting a spotlight on super polluters. Organizations like the Environment America Research and Policy Center and World Wide Fund for Nature now publish lists of the dirtiest power plants in the United States and Europe. Most recently, WattTime, an artificial intelligence firm with funding from Google.org, announced that it would track and publish the pollution of every power plant in the world, allowing users (including researchers, policy makers, and activists) to identify which super polluters are most active in real time. Other scholars have also joined the fray. In a recent global analysis of fossil-fuel- and biomass-burning power plants, Dan Tong and colleagues produced results consistent with those found in this book, arguing that "policies targeting a relatively small number of 'super-polluting' units could substantially reduce pollutant emissions and thus the related impacts on both human health and global climate."[43] Mary Collins, Ian Munoz, and Joseph JaJa[44] go a step further by showing how industrial hyperpolluters in the

United States disproportionately expose communities of color and low-income populations to harmful chemical emissions.

FINAL THOUGHTS

As the authors were preparing this book on CO_2 emissions for publication, another imperceptible menace that respects no borders was beginning to threaten the world community: the COVID-19 pandemic. In a broad sense, the pandemic reflected many of the challenges seen in efforts to address climate change, among them the lack of effective international coordination, the need for both systems-level responses and individual behavior change, and the disproportionate effect it is having on vulnerable populations. But while the early days of the pandemic saw oil prices plummet, pollution abate, and (at least according to social media) wildlife return to suddenly silent cities, the COVID-19 pandemic has only heightened our collective vulnerability, and to the extent vast resources are needed to prevent the future spread of the virus, many fear even fewer will be left to combat climate change. As economies recover and reboot, nations will need to develop highly focused and potent strategies if they are to have any chance of reversing the long-term acceleration in carbon pollution seen before the pandemic began.

While targeting hyperemitting power plants is not a panacea for climate change, the fact that we may now have just over a decade to bring it under control means that society must zero in on the worst of its polluters. Indeed, as Charles Perrow reminds us, climate change is ultimately a problem of organizations.[45] Climate scientists investigate the complex planetary trajectories and feedback loops of intermediate temperature rises; nation-states debate acceptable policy instruments that are amenable to their own economies, cultures, and polities; and protestors rail against various figures, groups, and mind-sets. But at the epicenter of catastrophic rising temperatures are the organizations that produce the energy on which entire social systems and economies depend. These organizations do not exist in a vacuum; rather, they are embedded in cultural, political, economic, and natural systems that shape their pollution outcomes, providing ample opportunities to increase emissions but also to abate them. We hope that society will build on this basic sociological observation—that organizations and their broader contexts *matter*—in deciding how to address its super polluters and other "villains" of climate change.

APPENDIX TO CHAPTER 2

DEPENDENT VARIABLE

The dependent variable in the regression analysis is national-level carbon emissions from fossil-fuel power plants for 2009. To create these national-level measures for the 161 nations in the study, we summed all plant-level emissions for the fossil-fuel power plants within each respective nation. As a validity check, we compared these summed values with the data gathered by the International Energy Agency (IEA) in its 2009 annual national measures of carbon dioxide (CO_2) emissions from fossil-fuel combustion for the electricity and heat production sector. These data are readily available online in the IEA's 2011 report on "CO_2 Emissions from Fuel Combustion." For the 121 nations that overlap between the two data sets (as data for 40 nations are missing in the IEA's data set), the Pearson's correlation for our summed measure of national carbon emissions from fossil-fuel power plants and the IEA's measure is .99, which validates our dependent variable.

ADDITIONAL INDEPENDENT VARIABLES

We included as additional independent variables national measures of population size, gross domestic product per capita (GDP per capita, measured in constant 2005 U.S. dollars), and trade as a percentage of gross domestic product. We obtained these data from the World Bank's World Development Indicators database. Population size and GDP per capita are the two most commonly assessed human drivers of national carbon emissions and together tend to explain a large amount of variation in overall emission levels across nations.[1] Generally speaking, more people usually means higher energy use, and the growth of national economies is also tightly connected to the burning of fossil fuels. Trade as a percentage of gross domestic product is a commonly used measure of nations' relative levels of integration in the world economy, and the effects of world economic integration have been the focus of much recent sociological research on the human drivers of national greenhouse gas emissions.[2] A common argument is that international trade allows more affluent nations to externalize their environmental impacts to some extent, which can lead to increases in emission levels for developing nations that manufacture products exported to, and consumed by, the populations of more affluent nations.[3]

We included as additional independent variables the number of fossil-fuel power plants within a nation, whether or not a nation is located in a tropical climate (coded 1 if a country's predominant latitude is less than 30 degrees from the equator), and the average price of electricity for each nation (measured in U.S. dollars). The data on average price of electricity were obtained from the IEA's Energy Prices and Taxes database. We also included measures of the percentages of a nation's fossil-fuel power plants whose primary fuels are coal, gas fossil fuels, and liquid fossil fuels. To create these national measures, we used data available from PLATTS, which identifies what the primary fuel source is for each plant.

MULTIPLE REGRESSION MODEL TECHNIQUES

To estimate the regression models, we used ordinary least squares (OLS) regression and robust regression, a combined approach suggested by researchers when analyzing cross-sectional data (data for one time point) for nations.[4] OLS is among the most common regression methods used across the social sciences. In the OLS models, we estimated jackknife standard errors, which require no assumptions about underlying distributions. Robust regression is a relatively conservative approach that downweights the influence of outliers in residuals.[5] In the robust regression models, we used a biweight tuning constant of 7, meaning seven times the median absolute deviation from the median residual, which is the default in Stata software. The results were very similar if we moderately increased or decreased the value of the biweight tuning constant.

TABLE A2.1 TOP FIVE POLLUTING PLANTS IN TOP TEN POLLUTING COUNTRIES, 2009

	Plant	Fuel	Parent Company	CO_2 (mT)
China	Waigaoqiao	Coal	Shenergy Company Ltd.	21,161,000
	Beilun	Coal	China Guodian Corp.	20,148,000
	Zouxian	Coal	China Huadian Group Corp.	18,527,000
	Huaneng Yuhuan	Coal	China Huaneng Group Corp.	17,797,000
	Ninghai	Coal	Shenhua Group Corp Ltd.	17,264,000
United States	Scherer	Coal	Southern Company	23,861,000
	James H. Miller Jr.	Coal	Southern Company	22,152,000
	Bowen	Coal	Southern Company	20,566,000
	W. A. Parish	Coal	NRG Energy	20,095,000
	Martin Lake	Coal	Energy Futures Holding Corp.	19,299,000
India	Vindhyachal	Coal	NTPC Ltd.	24,812,000
	Neyveli	Coal	Neyveli Lignite Corp. Ltd.	23,120,000
	Talcher Kaniha	Coal	NTPC Ltd.	21,227,000
	Ramagundam	Coal	NTPC Ltd.	19,417,000
	Korba STPS	Coal	NTPC Ltd.	16,172,000
Russia	Surgut-2	FGAS	E.On Ag	16,890,000
	Reftinskaya	Coal	Enel Spa	14,043,000
	Kostroma	FGAS	OGK-3 (Third Generation Co.)	13,063,000
	Surgut-1	FGAS	Gazprom	11,737,000
	Ryazan	FLIQ	Gazprom	9,401,700
Japan	Hekinan	FGAS	Chubu Electric Power Co.	18,664,000
	Futtsu	Coal	Tokyo Electric Power Co.	11,461,000
	Shinchi	Coal	Soma Kyodo Power Corp.	9,980,000
	Tachibanawan	Coal	J-Power	9,749,700
	Kawagoe	FGAS	Chubu Electric Power Co.	9,506,800

Germany	Niederaussem	Coal	RWE Ag	26,300,000
	Janschwalde	Coal	Vattenfall Group	23,600,000
	Weisweiler	Coal	RWE Ag	19,200,000
	Boxberg	Coal	Vattenfall Group	15,300,000
	Lippendorf	Coal	Vattenfall Group	12,800,000
South Korea	Poryong	Coal	Korea Midland Power	32,823,000
	Taean	Coal	Korea Western Power	30,347,000
	Dangjin	Coal	Korea East-West Power Co.	29,046,000
	Hadong	Coal	Korea Southern Power	28,719,000
	Samchonpp	Coal	Korea Southeast Power	24,113,000
Australia	Eraring	Coal	Origin Energy Ltd.	19,799,000
	Bayswater	Coal	Macquarie Generation	19,401,000
	Loy Yang A	Coal	AGL/TEPCO/RATCH/ MTAA Super	14,938,000
	Liddell	Coal	Macquarie Generation	13,399,000
	Gladstone	Coal	NRG Energy	11,140,000
United Kingdom	Drax	Coal	Drax Power Ltd.	20,500,000
	Cottam	Coal	Cheung Kong Infrastructure	8,590,000
	Ratcliffe	Coal	E.On Ag	7,610,000
	Longannet	Coal	Iberdola SA	7,390,000
	West Burton	Coal	Cheung Kong Infrastructure	7,260,000
Saudi Arabia	Al-Shuaibah Sec	FLIQ	Saudi Electricity Co.	19,691,000
	Rabigh	FLIQ	Saudi Electricity Co.	13,410,000
	Ghazlan	FGAS	Saudi Electricity Co.	12,335,000
	Al-Qurayah	FGAS	Saudi Electricity Co.	8,775,100
	Jubail IWPP	FGAS	Marafiq	8,398,400

TABLE A2.2 NATIONAL GINI COEFFICIENTS FOR DISPROPORTIONALITY IN PLANT-LEVEL CARBON EMISSIONS

Nation	Disproportionality Gini Coefficient	Number of Fossil-Fuel Power Plants	% Coal Fossil-Fuel Power Plants	% Gas Fossil-Fuel Power Plants	% Liquid Fossil-Fuel Power Plants
Afghanistan	38.47	32	0.00	9.38	90.63
Albania	14.52	4	0.00	0.00	100.00
Algeria	39.22	59	0.00	71.19	28.81
Antigua and Barbuda	24.50	6	0.00	0.00	100.00
Argentina	33.78	398	0.25	15.08	84.67
Armenia	5.48	2	0.00	100.00	0.00
Australia	41.53	434	9.45	47.24	43.32
Austria	38.20	78	5.13	73.08	21.80
Azerbaijan	32.97	16	0.00	87.50	12.50
Bahamas	30.83	30	0.00	0.00	100.00
Bahrain	21.61	12	0.00	100.00	0.00
Bangladesh	37.59	105	0.95	60.95	38.10
Barbados	18.72	4	0.00	0.00	100.00
Belarus	35.84	32	0.00	93.75	6.25
Belgium	37.75	196	5.61	71.43	22.96
Belize	23.68	8	0.00	0.00	100.00
Benin	36.39	12	0.00	0.00	100.00
Bermuda	17.31	3	0.00	0.00	100.00
Bolivia	30.11	16	0.00	50.00	50.00
Bosnia and Herzegovina	24.10	16	81.25	0.00	18.75
Brazil	52.61	157	3.82	31.21	64.97
Brunei Darus-salam	36.54	10	0.00	80.00	20.00
Bulgaria	41.93	27	40.74	59.26	0.00
Burkina Faso	29.50	41	0.00	0.00	100.00

Cambodia	27.72	32	0.00	0.00	100.00
Cameroon	28.13	10	0.00	0.00	100.00
Canada	50.91	449	4.23	45.43	50.33
Central African Republic	13.95	15	0.00	0.00	100.00
Chad	26.53	13	0.00	0.00	100.00
Chile	32.87	60	16.67	13.33	70.00
China	35.87	1,130	81.50	8.67	9.82
Colombia	36.20	65	9.23	43.08	47.69
Comoros	23.29	10	0.00	0.00	100.00
Congo, Dem. Rep.	15.28	35	0.00	2.86	97.14
Congo, Rep.	18.71	3	0.00	100.00	0.00
Costa Rica	27.69	7	0.00	0.00	100.00
Cote d'Ivoire	19.39	6	0.00	66.67	33.33
Croatia	33.16	18	5.56	27.78	66.67
Cuba	41.09	48	0.00	6.25	93.75
Cyprus	13.88	8	0.00	0.00	100.00
Czech Republic	42.55	74	44.60	54.05	1.35
Denmark	35.07	367	1.09	94.01	4.91
Djibouti	12.55	5	0.00	0.00	100.00
Dominican Republic	51.07	52	3.85	3.85	92.31
Ecuador	38.80	90	0.00	8.89	91.11
Egypt	31.73	78	0.00	38.46	61.54
El Salvador	34.45	12	0.00	0.00	100.00
Equatorial Guinea	4.85	2	0.00	0.00	100.00
Eritrea	14.93	6	0.00	0.00	100.00
Estonia	18.02	12	58.33	41.67	0.00
Ethiopia	34.19	44	0.00	0.00	100.00
Fiji	35.01	17	0.00	0.00	100.00
Finland	44.23	79	17.72	36.71	45.57
France	48.14	493	7.91	70.59	21.50

(*continued*)

TABLE A2.2 (*continued*)

Nation	Disproportionality Gini Coefficient	Number of Fossil-Fuel Power Plants	% Coal Fossil-Fuel Power Plants	% Gas Fossil-Fuel Power Plants	% Liquid Fossil-Fuel Power Plants
Gabon	37.91	31	0.00	9.68	90.32
Gambia	20.16	8	0.00	0.00	100.00
Georgia	11.62	6	0.00	66.67	33.33
Germany	58.17	965	12.23	71.71	16.06
Ghana	13.07	9	0.00	0.00	100.00
Greece	41.93	84	7.14	34.52	58.33
Greenland	39.36	21	0.00	0.00	100.00
Guatemala	31.27	15	6.67	0.00	93.33
Guinea	25.18	13	0.00	0.00	100.00
Guyana	26.02	19	0.00	0.00	100.00
Haiti	26.60	13	0.00	0.00	100.00
Honduras	30.07	26	0.00	0.00	100.00
Hong Kong	17.22	8	25.00	12.50	62.50
Hungary	52.29	68	8.82	82.35	8.82
India	46.97	737	37.72	18.86	43.42
Indonesia	58.15	533	6.75	12.57	80.68
Iran	38.51	140	0.00	45.00	55.00
Iraq	32.04	23	0.00	0.00	100.00
Ireland	40.25	38	2.63	78.95	18.42
Israel	33.46	25	12.00	36.00	52.00
Italy	53.81	536	3.92	65.49	30.60
Japan	42.23	1,908	4.61	38.63	56.76
Jordan	21.53	23	0.00	21.74	78.26
Kazakhstan	50.68	51	52.94	29.41	17.65
Kenya	24.72	21	0.00	0.00	100.00
Kiribati	23.11	4	0.00	0.00	100.00
Kuwait	30.41	19	0.00	26.32	73.68
Kyrgyz Republic	9.96	3	33.33	66.67	0.00

Latvia	19.78	29	0.00	82.76	17.24
Lebanon	26.13	18	0.00	0.00	100.00
Liberia	12.81	3	0.00	0.00	100.00
Libya	34.07	58	0.00	10.35	89.66
Lithuania	22.06	9	0.00	77.78	22.22
Luxembourg	11.29	57	0.00	100.00	0.00
Macau	7.19	4	0.00	0.00	100.00
Macedonia	15.38	6	33.33	33.33	33.33
Madagascar	39.29	116	0.00	0.00	100.00
Malawi	18.25	9	0.00	0.00	100.00
Malaysia	40.04	161	3.73	23.60	72.67
Maldives	53.22	189	0.00	0.00	100.00
Mali	36.22	21	0.00	0.00	100.00
Mauritania	21.71	27	0.00	0.00	100.00
Mauritius	24.65	9	33.33	0.00	66.67
Mexico	45.84	177	2.26	46.33	51.41
Moldova	31.17	13	0.00	100.00	0.00
Mongolia	22.11	25	32.00	0.00	68.00
Morocco	41.18	18	16.67	11.11	72.22
Mozambique	15.72	8	0.00	25.00	75.00
Namibia	2.85	3	33.33	0.00	66.67
Netherlands	48.79	591	0.51	98.14	1.35
New Zealand	38.11	47	6.38	70.21	23.40
Nicaragua	22.56	26	0.00	0.00	100.00
Niger	40.95	11	9.09	0.00	90.91
Nigeria	37.09	106	0.00	35.85	64.15
Norway	22.55	20	5.00	45.00	50.00
Oman	38.56	98	0.00	28.57	71.43
Pakistan	42.58	117	0.86	32.48	66.67
Panama	29.29	80	0.00	0.00	100.00
Papua New Guinea	25.20	34	0.00	5.88	94.12
Peru	37.43	285	0.35	3.86	95.79
Philippines	34.44	281	4.27	2.49	93.24

(*continued*)

TABLE A2.2 (*continued*)

Nation	Disproportionality Gini Coefficient	Number of Fossil-Fuel Power Plants	% Coal Fossil-Fuel Power Plants	% Gas Fossil-Fuel Power Plants	% Liquid Fossil-Fuel Power Plants
Poland	54.72	287	79.79	17.77	2.44
Portugal	44.82	190	1.05	45.79	53.16
Puerto Rico	33.62	20	5.00	5.00	90.00
Qatar	30.42	15	0.00	100.00	0.00
Romania	45.31	53	35.85	52.83	11.32
Russia	49.46	529	19.28	65.97	14.75
Saint Kitts and Nevis	12.09	3	0.00	0.00	100.00
Saint Lucia	9.40	3	0.00	0.00	100.00
Saint Vincent	16.67	6	0.00	0.00	100.00
Samoa	18.65	4	0.00	0.00	100.00
Sao Tome and Principe	4.82	4	0.00	0.00	100.00
Saudi Arabia	49.74	208	0.00	15.39	84.62
Senegal	39.66	53	0.00	0.00	100.00
Serbia	22.98	22	45.46	9.09	45.46
Sierra Leone	29.77	16	0.00	0.00	100.00
Singapore	20.90	40	0.00	37.50	62.50
Slovak Republic	41.70	36	41.67	30.56	27.78
Slovenia	21.82	23	21.74	30.44	47.83
Solomon Islands	33.94	11	0.00	0.00	100.00
South Korea	39.61	198	15.15	33.33	51.52
Spain	40.97	485	3.92	66.19	29.90
Sri Lanka	31.98	25	0.00	0.00	100.00
Sudan	30.86	37	0.00	0.00	100.00
Suriname	19.08	14	0.00	0.00	100.00
Sweden	34.77	49	6.12	44.90	48.98

Tanzania	9.30	37	2.70	8.11	89.19
Thailand	45.03	92	15.22	54.35	30.44
Togo	18.55	4	0.00	0.00	100.00
Tonga	9.89	4	0.00	0.00	100.00
Trinidad and Tobago	20.31	7	0.00	85.71	14.29
Tunisia	29.78	20	0.00	75.00	25.00
Turkey	38.05	304	15.79	51.65	32.57
Turkmenistan	35.06	10	0.00	100.00	0.00
Uganda	7.27	13	0.00	0.00	100.00
Ukraine	29.14	48	29.17	60.42	10.42
United Arab Emirates	33.83	71	0.00	70.42	29.58
United Kingdom	50.55	782	3.07	79.41	17.52
United States	48.86	2,612	21.75	53.52	24.73
Uruguay	14.38	3	0.00	0.00	100.00
Uzbekistan	30.14	10	20.00	70.00	10.00
Vanuatu	23.09	5	0.00	0.00	100.00
Venezuela	46.67	51	0.00	68.63	31.37
Vietnam	28.68	45	28.89	20.00	51.11
Yemen	33.92	31	0.00	0.00	100.00
Zambia	15.55	14	0.00	0.00	100.00
Zimbabwe	17.38	6	66.67	0.00	33.33

TABLE A2.3 ELASTICITY MODELS OF NATIONAL-LEVEL CO_2 EMISSIONS FROM FOSSIL-FUEL POWER PLANTS, 2009

	Model 1		Model 2		Model 3	
	OLS Jacknife	Robust Regression	OLS Jacknife	Robust Regression	OLS Jacknife	Robust Regression
Gini coefficient for disproportionality	.65***	.37**	.55**	.39**	.70***	.47**
	(.22)	(.21)	(.24)	(.23)	(.21)	(.20)
Population size	.93***	.99***	.70***	.78***	.80***	.85***
	(.05)	(.05)	(.10)	(.09)	(.05)	(.05)
GDP per capita	.99***	.96***	.71***	.70***	.88***	.86***
	(.07)	(.06)	(.11)	(.09)	(.07)	(.06)
Number of fossil-fuel power plants			.18	.10		
			(.13)	(.12)		
Tropical climate			−.08	−.07		
			(.12)	(.10)		
Price of electricity			−2.87**	−2.96**	−2.84**	−3.09**
			(1.54)	(1.67)	(1.55)	(1.67)

Percentage coal fossil-fuel power plants		.31***	.27***	.36***	.32***
		(.08)	(.07)	(.06)	(.07)
Percentage gas fossil-fuel plants		.08	.12		
		(.11)	(.09)		
Percentage liquid fossil-fuel plants		–.08	–.04		
		(.12)	(.11)		
Trade as percentage of GDP		.31	.33		
		(.22)	(.23)		
R^2	.79	.83		.82	

Notes: N = 161 nations; all variables except tropical climate are in base 10 logarithmic form; biweight tuning constant is 7 in robust regression models; since robust regression downweights the influence of cases based on the size of their error terms, the R^2 statistic cannot be calculated for such models. Standard errors are in parentheses.

$^{**}p < .05$

$^{***}p < .01$ (one-tailed test)

APPENDIX TO CHAPTER 3

DATA

The unit of analysis is the plant, and the constructed data set consists of 19,525 cases. To assess our three hypotheses, we used the international data set on fossil-fuel power plants' CO_2 emissions in 2004 and 2009 from the Center for Global Development's Carbon Monitoring for Action (CARMA) database and added indicators for the following for each plant: indicators of its nation's position in the world-system and proenvironmental world society, its nation's political-legal system, the plant's organizational characteristics, and other potentially relevant factors. CARMA assigns to each plant a unique S&P Global Platts identification code. This code enables a researcher to append additional information gathered by Platts and other sources on the organizational characteristics of each plant and its parent firm. CARMA also provides the address and coordinates of each plant, which can be used to append other data on its host country, including whether it is located in the core, semiperiphery, or periphery; its memberships in environmental international nongovernmental organizations (EINGOs) and treaties, and its political institutions.

DEPENDENT VARIABLES

Our first dependent variable, obtained from the CARMA data set, is carbon dioxide (CO_2) emission rate, or total kilograms of CO_2 emitted by a plant divided by the net megawatt-hours of electricity it generated in 2009. Because this measure of emissions is highly skewed, we used a logarithmic transformation (i.e., a logged value) of it in our analyses. We also examined whether the configurations we hypothesized as having a positive impact on emission rate have the same effect our second dependent variable, CO_2 emission level (i.e., total kilograms of CO_2 emitted by a plant in 2009), which was also logged due to skewness.

KEY INDEPENDENT VARIABLES

Our measure of world-system position is based on a principal components analysis with a varimax rotation of categorical and continuous measures used in past research. Using multiple indicators of a concept poses a challenge for fuzzy-set qualitative comparative analysis (fsQCA) because the number of possible combinations of conditions increases exponentially (2^k) as more conditions are included in the analysis, making the interpretation of configurations with numerous conditions unwieldy. Following the recommendations of Benoit Rihoux and Charles Ragin[1] and Peer Fiss,[2] we addressed this situation using data reduction techniques. Specifically, we conducted principal component factor analyses of various indicators of world-system position, world environmental norms, and organizational factors.

The world-systems position analysis included dummy variables for location in the core, semiperiphery, and periphery (taken from Rob Clark and Jason Beckfield's trichotomous country-level scheme[3]) as well as measures of individual nations' exports and foreign direct investment. As shown in table A3.1, the analysis produced a two-factor solution, with core and periphery loading highly on the first factor, which captured proximity to the world-system's core. Both items were combined in a scale that showed very good reliability (a = .782). Exports and foreign direct investment loaded on the second factor, but when they were combined in a scale, reliability was poor (a = .379). Therefore, rather than using this second scale in our fuzzy-set analyses, we incorporated exports and foreign direct investment in our regression models as controls. We also decided against including a set for core, semiperiphery, and periphery location in our fuzzy-set analyses; all of these variables are binary, and a plant obviously cannot be in more than one, so any configuration that involved all three would be eliminated from the analysis.

Our factor analysis of normative engagement (see table A3.2) yielded a one-factor solution that included five items used regularly in world society research: national memberships in EINGOs, comprehensive environmental law, environmental

TABLE A3.1 PRINCIPAL COMPONENT FACTOR ANALYSIS OF WORLD-SYSTEM POSITION

Survey Item	Factor 1	Factor 2
Core	.944	.027
Semiperiphery	–.040	–.001
Periphery	–.853	–.024
Exports	.257	.841
Foreign direct investment	–.162	.880
Eigenvalue	1.712	1.481
Portion of variance explained by Eigenvalue	.343	.296

TABLE A3.2 PRINCIPAL COMPONENT FACTOR ANALYSIS OF ENGAGEMENT WITH WORLD ENVIRONMENTAL NORMS

Survey Item	Factor 1
EINGOs	.701
Comprehensive environmental law	.652
Environmental impact legislation	.748
Environmental ministry	.724
Environmental treaties	.522
Eigenvalue	2.271
Portion of variance explained by Eigenvalue	.454

impact legislation, environmental ministry, and number of ratified international environmental treaties. These items were combined in a scale with high reliability (a = .720). Data came from multiple editions of *The Yearbook of International Organizations*, published annually by the Union of International Associations. The yearbook provides membership information for more than thirty-seven thousand international organizations. Longhofer and colleagues[4] used the yearbook's subject indices to generate a list of all EINGOs, organized by founding date. Coders then sampled every eighth organization (or roughly 13 percent of the population) and gathered annual membership information for each one. This yielded a final sample of fifty-four EINGOs. We thank David Frank for sharing these data with us. Data on environmental legislation were also used by Longhofer et al. and came from ECOLEX (www.ecolex.org). Data on treaties came from the United Nations Environment Programme's Environmental Data Explorer (www.geodata.grid.unep.ch) and were provided by Tricia Bromley.

To determine the effects of political checks and balances, we used an updated version of Mario Bergara, Witold Henisz, and Pablo Spiller's[5] index of political constraints (https://mgmt.wharton.upenn.edu/faculty/heniszpolcon/polcondataset/). This index was derived from three component variables: judicial effectiveness (e.g., median tenure of supreme or high court justices), formal constraints on executive discretion (e.g., the existence of federal units or a bicameral legislature elected under independent voting rules), and informal constraints on executive discretion (e.g., quality of the bureaucracy). High scores on this measure (which theoretically ranges from 0 to 1 but which actually peaks at .7) indicate that a nation's government more resembles a decentralized decision-making system and low scores a centralized one. This measure is related to but distinct from indices of democracy. While democracies tend to favor decentralized decision-making, this is not always the case, as, for example, where a unicameral parliament is governed by a majoritarian party. Also, we found that, unlike our measure of political institutions, traditional indicators of democracy (e.g., those identified by Freedom House) do not have significant direct or indirect effects on plants' emissions.

Our two organizational measures were derived from a factor analysis of several indicators of plant age, size, and ownership type. As shown in table A3.3, the analysis produced a two-factor solution with three items of size and ownership (percentage of a nation's electrical capacity owned by a plant's parent firm, percentage of a nation's electrical output produced by a plant's parent firm, and whether owned by a utility) loading on the first factor, which seems to measure whether a plant is part of a *dominant utility*, and with two items gauging plant age loading highly on the second factor. The reliabilities for the two resulting scales are, respectively, .803 and .961. One item, private ownership, did not load on either of the factors and was included as a control in the

TABLE A3.3 PRINCIPAL COMPONENT FACTOR ANALYSIS OF ORGANIZATIONAL FACTORS

Survey Item	Factor 1	Factor 2
Average generator age	.004	**.981**
Maximum generator age	.038	**.977**
Private ownership	.017	–.107
Utility ownership	**.483**	–.011
Share of electrical capacity	**.961**	.018
Share of electrical output	**.960**	.019
Eigenvalue 2	.083	.929
Portion of variance explained by Eigenvalue	.347	.322

regression analyses. Platt's World Electric Power Plants Database does not distinguish between publicly owned and investor-owned utilities. According to its codebook, there are too many varieties of utilities today to make a precise determination of what percentage of a facility is owned by whom (e.g., many state-owned power companies have been partially privatized, but the original government owner may retain a golden share). Still, as Charles Perrow suggests,[6] the ownership status of utilities may be less consequential for climate change than their size, as captured by our scale's other two size-related items.

CONTROLS

We controlled for whether a plant's fuel source is primarily coal (1 = coal) or primarily gas (1 = natural gas), its electrical capacity (nameplate capacity), and whether it is under private ownership

(1 = private independent power plant—this is a facility that was either newly built or acquired after having been controlled by an investor-owned or a state-owned utility and then operated in a nominally competitive fashion).

When testing the effects of configurations on plants' emission levels, we also included a control for capacity utilization rate (percentage of potential output that was produced). We did not include capacity utilization rate in our models for emission rates because including it and electrical capacity in the same equation essentially confounds the denominator of the dependent variable. Importantly, findings are substantively the same for the configurations if capacity utilization rate is included as a predictor of emission rates.

We also controlled for macrolevel factors shown in prior research to shape anthropogenic CO_2 emissions, including national-level measures of population size, wealth (gross domestic product [GDP] per capita), average price of electricity, whether a carbon tax has been adopted in a plant's country (1 = yes), exports (as a percentage of GDP), and foreign direct investment (as a percentage of GDP).[7] And to capture other conditions from the past that might influence plants' present environmental performance, we included indicators of our two emission measures for 2004 in some of the regression models.

Table A3.4 lists all included variables and their sources. When conducting our regression analyses (discussed shortly), all nonbinary independent variables and controls were logged to assess their multiplicative effects, a well-established practice in social science research on carbon emissions.[8] We measured all of our predictors in 2009 except for the following: prior emission rate, prior emission level, and our measure of proximity to the world-system's core zone, which includes items covering years up to 2000 and is arguably the best available estimate of the international trade regime during the period of study.

TABLE A3.4 VARIABLE SUMMARY

Variable	Source (Year)	Mean/S.D.
Dependent Variables		
CO_2 emission rate	CARMA (2009)	768.84/309.91
CO_2 emission level	CARMA (2009)	518639/1835012
Key Independent Variables		
Global		
Core (fuzzified)	Clark and Beckfield (2000)	.84/.35
Normative engagement (fuzzified)	Union of International Associations (2009); ECOLEX; United Nations Environment Programme	.51/.29
Political		
Political checks and balances (fuzzified)	Henisz (2002)	.47/.30
Organizational		
Dominant utility (fuzzified)	S&P Global Platts (2009)	.43/.26
Plant age (fuzzified)	S&P Global Platts (2009)	.47/.31
Controls		
Plant Characteristics		
Primarily coal (1 = coal)	S&P Global Platts (2009)	.14/.35
Primarily gas (1 = natural gas)	S&P Global Platts (2009)	.41/.49
Electrical capacity	S&P Global Platts (2009)	164.72/438.11
Capacity utilization rate	S&P Global Platts (2009)	.33/.24
Private ownership (1 = yes)	S&P Global Platts (2009)	.13/.34
Prior CO_2 emission rate	CARMA (2004)	740.75/401.96
Prior CO_2 emission level	CARMA (2004)	549,793/1,918,063
Country Characteristics		
Population	World Bank (2009)	2.06e + 08/3.52e + 08
Wealth (GDP per capita)	World Bank (2009)	22,947/17,094
Price of electricity (US cents/ kilowatt-hour)	Rosa and Dietz (2012)	.19/.08
Carbon tax (1 = yes)	International Energy Agency (2009)	.05/.22
Exports (% of GDP)	World Development Indicators (2009)	20.15/22.87
Foreign direct investment (% of GDP)	World Development Indicators (2009)	2.37/3.49

METHODS

The evaluation of our hypotheses required a mixed methods strategy. To assess our first hypothesis that global, political, and organizational factors combine in complex ways to create multiple pathways to high CO_2 emission rates, we needed a method capable of testing higher-order interactions. To assess our second and third hypotheses that these pathways affect emission rates and emission levels net of controls, we needed a method that can test the effects of higher-order interactions while also accounting for other potential causes.

fsQCA is especially well suited for evaluating higher-order interactions and determining which of several possible combinations of factors are most relevant for an outcome.[9] Regression techniques can evaluate interactions as well, but it is difficult to interpret anything beyond simple, two-way interactions within a regression format. Regression also presumes that a statistically significant interaction can be generalized to all cases under investigation when, in fact, it may be possible only in some cases. Multilevel regression, though, can do something that fsQCA cannot: it can test the effects of higher-order combinations net of other controls and capture fixed aspects of the units within which cases are situated. Next, we describe fsQCA and multilevel regression techniques and outline how we used them to test our hypotheses.

fsQCA

Qualitative comparative analysis (QCA) and its fuzzy-set variant, fsQCA, treat cases as combinations of attributes and use Boolean algebra to derive simplified expressions of combinations associated with an outcome.[10] For example, given an outcome set

Y and predictors A and B, QCA determines which combinations of A and B (i.e., AB, Ab, aB, ab) are most likely to produce Y. In a QCA framework, the term *set* is used instead of *variable* to stress the idea that each variable has been transformed to represent an individual case's level of membership in a given condition (e.g., a facility's membership in the set of organizations with high emission rates).

The combination of individual sets—for example, facilities that are located in the core zone *and* in countries with few political checks and balances—is then referred to as a *configuration*. Sets are labeled with uppercase and lowercase letters. When working with crisp sets, or sets that are all dichotomous indicators, uppercase letters signify 1 (fully in A) and lowercase letters signify 0 (fully out of A). When working with fuzzy sets, or sets that can take on a value between 0 and 1, uppercase letters indicate the level of set membership (e.g., the value of A), and lowercase letters indicate 1 minus the set membership (e.g., 1 - A). In the case of fsQCA, then, individual organizations can be more or less a member of a particular set (e.g., .33 would indicate something like "more out than in but still somewhat in" the set, whereas .75 would signify something like "more in than out but not entirely in" the set). While some variation is lost in "fuzzifying" measures, this limitation is more than offset by fsQCA's ability to assess higher-order interactions. Combining fuzzy sets into configurations is usually done using the minimum operator, so AB = min(A, B) or aB = min{(1 - A), B}. For example, a case with a fuzzy score of .6 on A and .3 on B would be said to have a fuzzy score of .3 in the configuration AB. Unlike variable-based methods that are founded on the notion of unifinality and seek to estimate a single recipe for all cases under examination, QCA methods explicitly take the idea of equifinality into account, allowing different subsets to produce the same outcome through different pathways.

The effect of a predictor can vary by type of case, as indicated by cases' joint values on multiple other predictors (i.e., statistical moderation). For example, predictors like location in the core of the world-system may differ based on other factors, which include their (combination of) values on other attributes. Therefore, QCA solutions can contain multiple pathways or configurations or recipes, with each pathway specifying the presence or absence of each predictor in the set.

In our fsQCA analyses, we converted the dependent and key independent variables into fuzzy scores because several of them cannot be easily categorized as full membership (1) or nonmembership (0) in a set, which would be required if using conventional QCA techniques. For example, it is not immediately obvious what constitutes a high CO_2 emission rate. fsQCA addresses this type of problem by having researchers recode their measures continuously as degrees of membership (or in the interval between 0 and 1) based on theoretical or substantive knowledge. Key to this coding procedure is deciding which cases are the most ambiguous or should be assigned a value of .5. Once variables are calibrated, fsQCA can then examine not only the level of overlap between independent variables but also the extent to which certain combinations of independent variables overlap or are a subset of the dependent variable (if X, then Y).

The fit of these solutions is described in terms of their consistency and coverage (see table 3.1). Consistency is a measure of how often the outcome is present when the conditions in a given pathway are met (the proportion of cases in X that are also in Y). A high consistency score (near 1) indicates that a solution term is near to being sufficient for an outcome. Coverage measures how often a certain pathway is present when the solution is present (the proportion of cases in Y that are also in X). A high coverage score indicates that a certain pathway is relevant—i.e., has empirical weight. Both consistency and coverage are calculated using the same formula, Σmin(Xi, Yi)/ΣYi. However, coverage is calculated only after it has been established that a given pathway (X) is a consistent subset of Y.[11]

Regulatory agencies like the U.S. Environmental Protection Agency do not specify the point beyond which a fossil-fuel power plant's CO_2 emission rate should be classified as high, but they do suggest that to screen out especially harmful plants, it is useful to compare a plant's emissions rate to the average for all fossil-fueled plants.[12] In keeping with this recommendation, therefore, we converted our dependent variable to a fuzzy score using the following algorithm, recommended by Ragin for mean-based factors:[13]

fuzzy score = exp(2*z_score)/(1 + exp(2*z_score))
where z_score = (raw_score - mean)/(standard deviation)

According to this formula, cases with scores closer to 0 are more out of a set, whereas cases with scores closer to 1 are more in a set. In calibrating our key independent variables, we used the same formula if no theory- or knowledge-based information existed about their membership properties. Table A3.5 reports that of the thirty-two hypothetically possible configurations ($2^5 = 32$), the one that was most often empirically observed (see the Frequency column) was plants that are in the core, are located in nations that are normatively engaged, have political checks and balances, are not owned by dominant utilities, and are old (CNPdA = 2,969; cases were determined to belong to a particular configuration using a best-fit measure).

Table A3.6 assesses the relationship between the key independent variable sets. Specifically, following the procedures recommended by Kyle Longest and Stephen Vaisey[14], it determines

TABLE A3.5 CONFIGURATIONS WITH THE GREATEST NUMBER OF CASES

Best Fit	Frequency	Percentage
BECs	1,797	8.95
BeCs	1,642	8.18
BecS	252	1.26
Becs	82	0.41
bECs	9,774	48.70
beCs	3,795	18.91
becS	1,538	7.66
becs	1,190	5.93

B = plant size
E = environmental international nongovernmental organizations
C = core
S = semiperiphery

which configurations have consistencies with high CO_2 emission rates (Y-consistency) significantly greater than .800 (see the table section Y-Consistency Versus Set Value) *and* significantly greater consistencies with high CO_2 emission rates (Y-consistency) than its negation (N-consistency) (see the table section Y-Consistency Versus N-Consistency). We set the threshold at .80 ("almost always sufficient"), which is considered appropriate for samples with a large N.[15] Contradictory configurations occur when cases in the same configuration show different outcomes, weakening set theoretic consistency and making it more difficult to draw inferences about causal relationships. Stata reports only figures regarding the "Y-Consistency Versus N-Consistency" test for common sets that passed as true (indicated under the column "Common Sets" as "Y").

Based on these tests, the most consistent configurations with high CO_2 emission rates (see configurations marked as "Y" under

TABLE A3.6 CONSISTENCIES OF CONFIGURATIONS

Y-Consistency Versus N-Consistency					
Set	**YCons**	**NCons**	**F**	**P**	**NumBestFit**
Becs	0.972	0.928	23.59	0.000	82
BecS	0.955	0.863	93.82	0.000	252
BeCs	0.981	0.796	1,777.79	0.000	1,642
BECs	0.950	0.848	890.30	0.000	1,797
Y-Consistency Versus Set Value					
Set	**YCons**	**Set Value**	**F**	**P**	**NumBestFit**
beCs	0.763	0.800	1,428.30	0.000	3,795
Becs	0.972	0.800	5,877.06	0.000	82
BecS	0.955	0.800	5,977.61	0.000	252
BeCs	0.981	0.800	11,2391.10	0.000	1,642
BECs	0.950	0.800	30,287.51	0.000	1,797
Common sets (initial configurations that passed as true): Becs BecS BeCs BECs					
Minimum configuration reduction set: Bec BC					

B = plant size
E = environmental international nongovernmental organizations
C = core
S = semiperiphery

"Common Sets") are (1) plants that are in core nations that are normatively disengaged and have few political checks and balances and that are not owned by dominant utilities and are not old (Cnpda), (2) plants that are in core nations that are normatively disengaged and have few political checks and balances and that are owned by dominant utilities and are not old (CnpDa), (3) plants that are in core nations that are normatively disengaged and have political checks and balances and that are owned by dominant utilities and are not old (CnPDa), (4) plants that are in core nations that are normatively disengaged and have few political checks and balances and that are not owned by dominant utilities and are old (CnpdA), (5) plants that are in core nations that are normatively disengaged and have political checks and balances and that are not owned by dominant utilities and are old (CnPdA), (6) plants that are in noncore nations that are normatively disengaged and have few political checks and balances and that are owned by dominant utilities and are old (cnpDA), (7) plants that are in noncore nations that are normatively engaged and have few political checks and balances and that are owned by dominant utilities and are old (cNpDA), (8) plants that are in core nations that are normatively disengaged and have few political checks and balances and that are owned by dominant utilities and are old (CnpDA), (9) plants that are in core nations that are normatively disengaged and have political checks and balances and that are owned by dominant utilities and are old (CnPDA), (10) plants that are in core nations that are normatively engaged and have few political checks and balances and that are owned by dominant utilities and are old (CNpDA), and (11) plants that are in core nations that are normatively engaged and have political checks and balances and that are owned by dominant utilities and are old (CNDPA).

Because these configurations may logically overlap, they were reduced, collapsing the eleven initial configurations into five. This process, also known as minimization, is done by an automated algorithm that takes two configurations at a time and determines whether they have the same outcome. If they do and if their configurations are different with respect to only one condition, this is deemed not to be an important causal factor, and the two configurations are collapsed into one. This process of comparisons is continued, looking at all configurations, including newly collapsed ones, until no further reductions are possible. This minimum configuration reduction set turned out to be the final reduction set, which we discuss in the "Results" section of chapter 3.

MULTILEVEL REGRESSION

Our data structure was such that power plants are cross-nested within countries and power companies. We accounted for this nesting when conducting our regression analysis of power plants' CO_2 emissions for 2009 by using a hierarchical linear model with two random intercepts (one for countries and the other for power companies).[16] Our model also incorporated an unbalanced design that accounts for the fact that each power company does not have the same number of plants. We assessed model fit penalized for the number of estimated parameters using Bayesian information criterion statistics, which are available in Stata.

Whereas in the past fsQCA was viewed as a stand-alone alternative to regression analysis, today it is increasingly integrated with regression in mixed methods approaches, especially when analyzing large samples like ours.[17] One common strategy is to use fsQCA to assess complex interaction effects and then employ regression to test the robustness of the causally relevant configurations identified by fsQCA.[18] This can be done, according to Charles Ragin[19] and Peer Fiss, Dmitry Sharapov, and Lasse Cronqvist,[20] by entering configurations' membership scores as predictors in a regression equation that includes controls.

Accordingly, we used this strategy to test our second and third hypotheses. To assess the hypothesis that there is no single structural pathway to emission rates, we conducted an fsQCA analysis to determine whether plants' global, political, and organizational characteristics combine to create the four distinct pathways we predicted. (In other preliminary analyses not reported, we found that each of our key independent variables exerts significant, independent effects.) We also checked these configurations' robustness by including the solution membership of each as a predictor in our regression models containing controls (these predictors are actual membership scores as opposed to dummy variables for cases above or below a certain membership level), along with an interaction term for all of the conditions included in the configurations. The latter, according to Ragin,[21] ensures that cases that are doubly determined (from a logical point of view) are not doubly counted in a regression analysis. We did not split the data set and use one fraction for fsQCA and the other for the regression models for two reasons. First, the logic of generalizing from random samples to populations in regression analyses presupposes a substantial degree of homogeneity of cases in the population that may not exist. Second, random sampling is not appropriate for studies like this one that are interested in exploring the diversity of cases. A random sample may not represent the full diversity of cases, since some may occur only rarely in a population.[22] For the sake of parsimony, we did not present the interaction in the regression tables. And to assess the hypothesis that the four pathways increase emission levels, we reestimated our models using emission levels as the dependent variable.

APPENDIX TO CHAPTER 4

DATA FOR ANALYSIS OF U.S. POWER PLANTS

To analyze the environmental performance of U.S. utilities, we constructed a data set that includes indicators of U.S. fossil-fuel electric power generation facilities' carbon dioxide (CO_2) emissions in 2010 (North American Industry Classification System Code 221112) as well as other relevant factors. The unit of analysis is the power plant, and the analyzed data set consists of 1,129 cases. In this study, we examined the determinants of plants' emission rates and levels in 2010 and changes in those outcomes between 2005 and 2010 by controlling for rates and levels in 2005. We used a single cross section of 2010 data rather than continuous panels because the data from the Greenhouse Gas Reporting Program (GHGRP), administered by the U.S. Environmental Protection Agency (EPA), were available only for 2010 at the time this study was originally conducted and several of our predictors were measured only in that year.

Our sample (N = 1,129) contains only about a third of all power plants in the United States in 2010 (N = 3,406), largely because the GHGRP data on emissions primarily include plants that meet the EPA's criterion of a "major source" polluter (one that emits 25,000 metric tons or more of CO_2 equivalent in a year) and are required to submit emissions reports (N = 1,426). Of these plants, we excluded 297 from our analysis because information on their internal characteristics (e.g., type of ownership) and/or 2005 emissions was unavailable. Importantly, the 1,129 plants examined here, by themselves, account for 90.1 percent of all CO_2 emitted by the electricity sector in the United States.

DEPENDENT VARIABLES FOR ANALYSIS OF U.S. POWER PLANTS

Our two measures of emission outcomes—emission rate and emission level—are taken from the GHGRP, which began requiring power plants to submit information on their carbon pollution in 2010. Emission rate is operationalized as pounds of CO_2 released by a plant per kilowatt-hour (kWh) of electricity generated. Emission level is total pounds of CO_2 emitted by a plant. Because the latter variable is highly skewed, we used a logarithmic transformation of it in our analyses. To assess the determinants of plants' emissions over time, we included lagged measures of these variables for 2005 in our models. These measures effectively capture other conditions from the past that might influence plants' present environmental performance. They were constructed by aggregating generator-specific data gathered by the U.S. Energy Information Administration in 2005 to the plant level. Importantly, our measures of emission rates and emission levels are only weakly correlated (.118), suggesting not only that they might be determined by different factors but also that some factors like efficiency could have opposite effects on emission rates and levels.

KEY INDEPENDENT VARIABLE FOR ANALYSIS OF U.S. POWER PLANTS

Our key independent variable is heat rate, which is operationalized as the amount of energy used to generate 1 kWh of electricity. This measure is expressed in British thermal units per net kWh produced. Plants that score higher on this indicator do a less efficient job of capturing heat energy, and vice versa. The overall efficiency of a power plant encompasses the efficiency of numerous components of its generators. Minimizing heat losses, however, is regarded as the single greatest factor affecting a fossil-fuel power plant's efficiency.

CONTROLS FOR ANALYSIS OF U.S. POWER PLANTS

To ensure that the effects of heat rate are not artifacts of other relevant factors, we controlled for several characteristics of plants. Specifically, we tested the effects of plants' size (nameplate capacity measured in megawatts), whether coal is their primary fuel (1 = yes), and their median fuel price. In other analyses not reported here, we tested several other factors, including whether plants use equipment to control other pollutants such as nitrogen oxides (1 = yes), whether they either fall into a balancing authority area governed by an independent system operator or are part of a regional transmission organization that facilitates more efficient power flows and transactions (1 = yes), whether they are a public utility (1 = yes), whether there is coal industry influence (coal workers per 1,000 residents), whether there is oil and gas industry influence (oil and gas workers per 1,000 residents), whether there is Democratic control (1 = Democratic governor and Democratic majority in both chambers of the legislature), the percentage of state expenditures devoted to energy efficiency, the technical potential for renewable energy, the change in natural gas prices between 2005 and 2010, and the change in their census region's net electric output between 2005 and 2010 (a proxy for increases in electricity demand that might drive up emissions). We found these factors to be largely inconsequential.

ANALYTICAL STRATEGY FOR ANALYSIS OF U.S. POWER PLANTS

In conducting ordinary least squares regression analyses of the determinants of power plants' CO_2 emissions, we effectively controlled for the average differences across parent companies in any observable or unobservable predictors by including a dummy variable for each parent company in our models. In doing so, we also accounted for the fact that each company does not have the same number of plants.

DATA FOR ANALYSIS OF INTERNATIONAL POWER PLANTS

The primary data source for our international analysis of plants was the Center for Global Development's Carbon Monitoring for Action (CARMA) database. CARMA draws on three data sets: plant-level emission reports from the United States, the European Union, Canada, and India; plant- and company-level data from S&P Global Platts World Electric Power Plants Database; and country-specific power production data from the U.S. Energy Information Agency.[1] For nonreporting plants,

CARMA estimates emissions using a statistical model fitted to data from the reporting plants and detailed data from the other two sources of plant-level engineering specifications. Thus, CARMA provides information on the environmental performance of every country's power plants. According to Wheeler and Ummel, the creators of CARMA, for any given plant in the overall database, it is estimated that the reported value for CO_2 emission levels is within 20 percent of the actual value in 75 percent of the cases.

CARMA assigns to each plant a unique Platts identification code. This code enables a researcher to append additional information gathered by Platts and other sources on the organizational characteristics of each plant. CARMA also provides the address and coordinates of each plant, which can be used to append other data on its host country. Coordinates can be used in conjunction with geographic information system technology to map the location of a particular plant.

Our unit of analysis is the power plant, and the analyzed data set consists of 19,525 cases. We chose this unit of analysis because power plants are the sites where electricity is produced and roughly a quarter of all anthropogenic CO_2 emissions occur. Any study of the effects of power plants' technical and organizational features on emissions using more aggregated data would be subject to misinterpretation and the ecological fallacy.

DEPENDENT VARIABLE FOR ANALYSIS OF INTERNATIONAL POWER PLANTS

Our dependent variable is CO_2 emissions, or total pounds of CO_2 emitted by a plant in 2009. Because the variable is highly skewed, we used a logarithmic transformation of it in our analysis. To assess the determinants of plants' emissions over time, we included a lagged indicator of this variable for 2004 (logged) in our models. This effectively captures other conditions from the past that might influence plants' present environmental performance. As noted above, data for plants' emissions came from CARMA.

KEY INDEPENDENT VARIABLES FOR ANALYSIS OF INTERNATIONAL POWER PLANTS

We examined the effects of a power plant's thermal efficiency, which is operationalized as total energy produced as a percentage of heat energy generated. It is essentially the inverse of heat rate. We also examined the effects of plant age (year of oldest generator) and plant size (full nameplate capacity). These plant-specific measures came from Platts. We employed Rob Clark and Jason Beckfield's[2] trichotomous national-level measure of world-system position. Specifically, we used dummy variables to indicate whether a plant's nation is located in the core region (1 = yes, 0 = periphery) or the semiperiphery (1 = yes, 0 = periphery). And to gauge the degree to which a plant's nation is embedded in the world society, we used the (logged) number of memberships in environmental international nongovernmental organizations (EINGOs); supplemental analyses that instead used a per-capita measure of EINGOs yielded results that are consistent with the reported findings.[3] We included interactions between thermal efficiency and (1) plant age, (2) plant size, (3) core, and (4) EINGOs.

CONTROLS FOR ANALYSIS OF INTERNATIONAL POWER PLANTS

We controlled for a plant's potential to emit carbon pollution using indicators of its primary fuel (1 = coal) and capacity utilization rate, provided by Platts. The carbon content of coal varies by its moisture, but systematic information on this is not available for most plants and cannot be readily estimated. We also controlled for country-level factors known to increase consumer demand for electricity, including (logged) population, wealth (logged gross national product per capita), and average price of electricity.[4] These data were obtained from the World Bank's World Development Indicators database and the International Energy Agency.

In unreported sensitivity analyses, we also controlled for climatic conditions, using a dummy variable for tropical weather (coded 1 if a country's predominant latitude is less than 30 degrees from the equator).[5] The effect of this measure was not statistically significant, and its inclusion did not alter any of the reported findings.

ANALYTICAL STRATEGY FOR ANALYSIS OF INTERNATIONAL POWER PLANTS

We conducted multilevel regression analyses of the determinants of power plants' CO_2 emission levels using a fixed effects model that includes a dummy variable for each plant's parent company. This modeling procedure effectively controlled for the average differences across parent companies in any observable or unobservable predictors and uses an unbalanced design that accounted for the fact that each company does not have the same number of plants. Hausman tests indicated that a fixed effects model is more appropriate than a random effects model. Additional robustness checks showed that our standard error estimates are not biased by heteroskedasticity. In unreported models, we found that the effects of our predictors are essentially the same when using a change score (instead of a lagged dependent variable) specification, suggesting that our reported analysis is not substantially compromised due to the unavailability of longitudinal data.

APPENDIX TO CHAPTER 5

DATA

For both analyses, we constructed a data set that includes indicators of U.S. fossil-fuel electric power generation facilities' carbon dioxide (CO_2) emissions in 2010 (North American Industry Classification System [NAICS] Code 221112) as well as other relevant factors. The unit of analysis is the power plant, and the analyzed data set consists of 1,129 cases. In this study, we examined the determinants of plants' emission rates and levels in 2010 and changes in those outcomes between 2005 and 2010 by controlling for rates and levels in 2005. We used a single cross section of 2010 data rather than continuous panels because the data from the Greenhouse Gas Reporting Program (GHGRP), administered by the U.S. Environmental Protection Agency (EPA), were available only for 2010 at the time this study was originally conducted and several of our predictors were measured only in that year.

Our sample (N = 1,129) contains only about a third of all power plants in the United States in 2010 (N = 3,406), largely because the GHGRP data on emissions primarily include plants that meet the EPA's criterion of a "major source" polluter (one that emits 25,000 metric tons or more of CO_2 equivalent in a year) and are required to submit emissions reports (N = 1,426). Of these plants, we excluded 297 from our analysis because information on their internal characteristics (e.g., type of ownership) and/or 2005 emissions was unavailable. Importantly, the 1,129 plants examined here, by themselves, account for 90.1 percent of all CO_2 emitted by the electricity sector in the United States.

DEPENDENT VARIABLES

Our two measures of emission outcomes—emission rate and emission level—are taken from the GHGRP, which began requiring power plants to submit information on their carbon pollution in 2010. Emission rate is operationalized as the pounds of CO_2 released by a plant per kilowatt-hour (kWh) of electricity generated. Emission level is total pounds of CO_2 emitted by a plant. Because the latter variable is highly skewed, we used a logarithmic transformation of it in our analyses. To assess the determinants of plants' emissions over time, we included lagged measures of these variables for 2005 in our models. These measures effectively capture other conditions from the past that might influence plants' present environmental performance. They were constructed by aggregating generator-specific data gathered by the U.S. Energy Information Administration (EIA) in 2005 to the plant level.

KEY INDEPENDENT VARIABLES

Our indicators of states' climate-focused policies and energy policies with climate implications came from the Pew Center on

Global Climate Change and the Database of State Incentives for Renewables and Efficiency. Because the amount of experience a state has with a policy may affect its success, we tested the effects of each of these policies using a set of dichotomous variables—one indicating whether a particular state policy had been implemented for five years or more as of 2010 (1 = yes) and the other indicating whether a policy had been in place for one to four years (1 = yes). (We coded states that adopted a policy in 2010 as having one year of experience, those that adopted a policy in 2009 as having two years of experience, and so forth.) The comparison group consisted of plants in states that had never adopted a policy as of 2010 (0 = never).

Our other key independent variable—environmental nongovernmental organizations (ENGOs)—is expressed as the number of environment and conservation nonprofit organizations (NAICS Code 813312) in a plant's county and is taken from the Quarterly Census of Employment and Wages (https://www.bls.gov/cew/) conducted by the U.S. Bureau of Labor Statistics. We used this measure because it provides better granularity than state-level measures of environmental activism used in previous studies.[1] However, this measure is highly skewed, so we transformed it when conducting our regression analyses by taking its natural logarithm; to avoid simultaneity bias, we used the measure from 2009. This method is in keeping with previous studies that have examined the effects of county-level civic engagement on environmental, poverty, and health outcomes.[2] An even more fine-grained measure of ENGOs' presence would be data from the city level; however, given the large number of cities that are near power plants, we could not collect those data, and we instead had to rely on the county-level measures.

CONTROLS

Using data collected by the EIA, we controlled for (depending on the model specification) the effects of plants' characteristics—namely, whether coal is their primary fuel (1 = yes), their size (nameplate capacity measured in megawatts), the year they were founded, whether they use equipment to control other pollutants such as nitrogen oxides (1 = yes), whether they either fall into a balancing authority area governed by an independent system operator or are part of a regional transmission organization that facilitates more efficient power flows and transactions (1 = yes), and whether they are a public utility (1 = yes). With respect to pollution control equipment, we examined devices for pollutants other than nitrous oxide but found they have no effect on CO_2 emissions. Also, using data from the U.S. Statistical Abstracts, the U.S. Department of Energy's National Renewable Energy Laboratory, the American Council for an Energy-Efficient Economy, and the EIA, we assessed the effects of the following attributes of a plant's state and region: coal industry influence (coal workers per 1,000 residents), oil and gas industry influence (oil and gas workers per 1,000 residents), whether there is Democratic control (1 = Democratic governor and Democratic majority in both chambers of the legislature), the percentage of state expenditures devoted to energy efficiency, the technical potential for renewable energy, the population density (county), the median income (county), whether the plant is located in the south central or mountains/plains region where the largest gas fields by proved reserves are concentrated (1 = yes), the change in natural gas prices between 2005 and 2010, the change in its census region's net electric output between 2005 and 2010 (a proxy for increases in electricity demand that might drive up emissions), and the number of other tested policies (i.e., the total of climate-focused and climate-implication policies in a state sans the specific policy being examined).

ANALYTICAL STRATEGY

In conducting ordinary least square regression analyses of the determinants of power plants' CO_2 emissions, we effectively controlled for the average differences across parent companies in any observable or unobservable predictors by including a dummy variable for each parent company in our models. In doing so, we accounted for the fact that each company does not have the same number of plants. We also conducted robustness checks, the results of which indicated that our standard error estimates are not biased by heteroskedasticity. In unreported models, we found that the effects of our predictors are essentially the same when using a change score (instead of a lagged dependent variable) specification, suggesting that our reported analyses are not substantially compromised due to the unavailability of longitudinal data.

NOTES

1. WHO IS RESPONSIBLE FOR THIS MESS?: THE CLIMATE CRISIS AND HYPEREMITTING POWER PLANTS

1. "Taichung Power Plant World's Worst Polluter: Survey," *Taipei Times*, September 4, 2008, 11.
2. International Energy Agency (IEA), *Global Energy and CO_2 Status Report 2018* (Paris: IEA, 2019).
3. International Energy Agency, *Global Energy and CO_2 Status Report 2018*.
4. "Central Taiwan Power Plant Fined Again for Exceeding Coal Use," *Taiwan News*, December 14, 2019.
5. David Roberts, "A Major Utility Is Moving Toward 100 Percent Clean Energy Faster Than Expected," *Vox*, May 29, 2019. https://www.vox.com/energy-and-environment/2018/12/5/18126920/xcel-energy-100-percent-clean-carbon-free.
6. Steven Bernstein and Matthew Hoffman, "Climate Politics, Metaphors and the Fractal Carbon Trap," *Nature Climate Change* 9 (2019): 919–925.
7. Thomas Dietz, "Drivers of Human Stress on the Environment in the Twenty-First Century," *Annual Review of Environment and Resources* 42 (2017): 189–213; Andrew Jorgenson et al., "Social Science Perspectives on Drivers of and Responses to Global Climate Change," *Wiley Interdisciplinary Reviews: Climate Change* 10, no. 1 (2019): e554, https://doi.org/10.1002/wcc.554; Eugene Rosa and Thomas Dietz, "Human Drivers of National Greenhouse Gas Emissions," *Nature Climate Change* 2 (2012):

581–586; Kimberly Thomas et al., "Explaining Differential Vulnerability to Climate Change: A Social Science Review," *Wiley Interdisciplinary Reviews: Climate Change* 10, no. 2 (2019): e565, https://doi.org/10.1002/wcc.565.

8. David Victor, *Global Warming Gridlock: Creating More Effective Strategies for Protecting the Planet* (New York: Cambridge University Press, 2011).
9. The term *wicked* has its origins in the realization that social and environmental problems are too complex to yield to the simple kinds of systems analysis that were popular in the 1960s. C. West Churchman, "Guest Editorial: Wicked Problems," *Management Science* 14, no. 4 (1967): B141–B142.
10. Will Steffen et al., "Trajectories of the Earth System in the Anthropocene," *Proceedings of the National Academy of Sciences* 115, no. 33 (2018): 8252–8259.
11. Robert Pollin, "Advancing a Viable Global Climate Stabilization Project: Degrowth Versus the Green New Deal," *Review of Radical Political Economics* 51 (2019): 311–319; Juliet Schor and Andrew Jorgenson, "Is It Too Late for Growth?" *Review of Radical Political Economics* 51 (2019): 320–329.
12. Nicholas Stern, "Current Climate Models Are Grossly Misleading," *Nature* 530 (2016): 407–409; Martin L. Weitzman, "Fat-Tailed Uncertainty in the Economics of Catastrophic Climate Change," *Review of Environmental Economics and Policy* 5, no. 2 (2011): 275–292; Martin L. Weitzman, "A Review of William Nordhaus' *The Climate Casino: Risk, Uncertainty, and Economics for a Warming World*," *Review of Environmental Economics and Policy* 9, no. 1 (2015): 145–156.
13. Intergovernmental Panel on Climate Change (IPCC), *Climate Change 2014: Synthesis Report* (Geneva: IPCC, 2014): 422, https://www.ipcc.ch/report/ar5/wg2/.
14. Richard Revesz et al., "Global Warming: Improve Economic Models of Climate Change," *Nature* 508 (2014): 173–175.
15. Benjamin Sovacool, "What Are We Doing Here?: Analyzing Fifteen Years of Energy Scholarship and Proposing a Social Science Research Agenda," *Energy Research and Social Science* 1 (2014): 1–29.
16. William Freudenburg, "Privileged Access, Privileged Accounts: Toward a Socially Structured Theory of Resources and Discourses," *Social Forces* 84 (2005): 89–114.

17. Thomas Dietz, "Context Matters: Eugene A. Rosa's Lessons for Structural Human Ecology," in *Structural Human Ecology: New Essays in Risk, Energy, and Sustainability*, ed. Thomas Dietz and Andrew Jorgenson (Pullman: Washington State University Press, 2013), 189–215.
18. It also tends to ignore how policies to induce technological innovation may take too long to achieve ambitious near- and long-term emissions reduction targets. See Thomas Dietz et al., "Household Actions Can Provide a Behavioral Wedge to Rapidly Reduce U.S. Carbon Emissions," *Proceedings of the National Academy of Sciences* 106 (2009): 18452–18456.
19. Kenneth Gillingham, David Rapson, and Gernot Wagner, "The Rebound Effect and Energy Efficiency Policy," *Review of Environmental Economics and Policy* 10, no. 1 (2016): 68–88.
20. See also Political Economy Research Institute, "Toxic 100 Air Polluter Index (2018 Report Based on 2015 Data)," University of Massachusetts, Amherst, 2018, https://www.peri.umass.edu/toxic-100-air-polluters-index-2018-report-based-on-2015-data.
21. B. Ekwurzel et al., "The Rise in Global Atmospheric CO_2, Surface Temperature, and Sea Level from Emissions Traced to Major Carbon Producers," *Climatic Change* 144 (2017): 579–590.
22. International Energy Agency, *Global Energy and CO_2 Status Report 2018*; Corrine Le Quiere et al., "Global Carbon Budget 2018," *Earth Systems Science Data* 10 (2018): 2141–2194; Robert Jackson et al., "Global Energy Growth Is Outpacing Decarbonization," *Environmental Research Letters* 13, no. 12 (2018), https://doi.org/10.1088/1748-9326/aaf303.
23. John Urry, "The Problem of Energy," *Theory, Culture and Society* 31, no. 5 (2014): 3–20.
24. Quoted in G. Kirk, *Schumacher on Energy* (London: Jonathan Cape, 1982), 1–2.
25. Timmons Roberts and Bradley Parks, *A Climate of Injustice: Global Inequality, North-South Politics, and Climate Policy* (Cambridge, MA: MIT Press, 2007).
26. Stephen Bunker and Paul Ciccantell, *Globalization and the Race for Resources* (Baltimore: Johns Hopkins University Press, 2005).
27. Karl Marx and Friedrich Engels, *The Manifesto of the Communist Party* (Moscow: Foreign Languages, 1888 [1848]).
28. Jared Diamond, *Collapse: How Societies Choose to Fail or Succeed* (New York: Viking Press, 2005).

29. J. R. McNeill, *Something New Under the Sun: An Environmental History of the Twentieth-Century World* (New York: Norton, 2000).
30. Stephen Hilgartner and Charles L. Bosk, "The Rise and Fall of Social Problems: A Public Arenas Model," *American Journal of Sociology* 94, no. 1 (1988): 53–78.
31. Matthew Desmond, *Evicted: Poverty and Profit in the American City* (New York: Broadway Books, 2016).
32. Alice Goffman, *On the Run: Fugitive Life in an American City* (New York: Picador, 2015).
33. Elizabeth Armstrong and Laura Hamilton, *Paying for the Party: How College Maintains Inequality* (Cambridge, MA: Harvard University Press, 2013).
34. See Jeremy Brecher, *Climate Insurgency: A Strategy for Survival* (New York: Routledge, 2015); Christopher Chase-Dunn, *Global Formation: Structures of the World Economy* (Lanham, MD: Rowman & Littlefield, 1998).
35. Charles Perrow, "Organizations and Global Warming," in *Routledge Handbook of Climate Change and Society*, ed. Constance Lever-Tracy (New York: Routledge, 2010), 59–77.
36. William Catton and Riley Dunlap, "Environmental Sociology: A New Paradigm," *American Sociologist* 13 (1978): 41–49; Frederick Buttel, "New Directions in Environmental Sociology," *Annual Review of Sociology* 13 (1987): 465–488; David Pellow and Hollie Brehm, "An Environmental Sociology for the Twenty-First Century," *Annual Review of Sociology* 39 (2013): 229–250; cf. Eugene Rosa and Lauren Richter, "Durkheim on the Environment: Ex Libris or Ex Cathedra? Introduction to Inaugural Lecture to a Course in Social Science, 1887–1888," *Organization and Environment* 21, no. 2 (2008): 182–187.
37. See also Debra Javeline, "The Most Important Topic Political Scientists Are Not Studying: Adapting to Climate Change," *Perspectives on Politics* 12, no. 2 (2014): 420–434.
38. John Bellamy Foster, "Marx's Theory of Metabolic Rift," *American Journal of Sociology* 105, no. 2 (1999): 366–405.
39. John Bellamy Foster and Hannah Holleman, "Weber and the Environment: Classical Foundations for a Postmaterialist Sociology," *American Journal of Sociology* 117, no. 6 (2012): 1665; See also Brett Clark and Richard York, "Carbon Metabolism: Global Capitalism, Climate Change, and the Biospheric Rift," *Theory and Society* 34, no. 4 (2005): 391–428.

40. Fred Cottrell, *Energy and Society: The Relation Between Energy, Social Change, and Economic Development* (New York: McGraw-Hill, 1955).
41. Ryan Gunderson, "Explaining Technological Impacts Without Determinism: Fred Cottrell's Sociology of Technology and Energy," *Energy Research and Social Science* 42 (2018): 127–133.
42. Riley Dunlap and Robert J. Brulle, eds., *Climate Change and Society: Sociological Perspectives* (New York: Oxford University Press, 2015); Jorgenson at al., "Social Science Perspectives on Drivers of and Responses to Global Climate Change"; Rosa and Dietz, "Human Drivers of National Greenhouse Gas Emissions."
43. Thomas Burns, Byron Davis, and Edward Kick, "Position in the World-System and National Emissions of Greenhouse Gases," *Journal of World-Systems Research* 3, no. 3 (1997): 432–466; Roberts and Parks, *A Climate of Injustice.*
44. Ann Hironaka, *Greening the Globe: World Society and Environmental Change* (New York: Cambridge University Press, 2014); Kirsten Shorette et al., "World Society and the Natural Environment," *Sociology Compass* 11, no. 10 (2017), https://doi.org/10.1111/soc4.12511.
45. Dana Fisher et al., "Polarizing Climate Politics in America," *Research in Political Sociology* 25 (2018): 1–23; Rachael Shwom, "A Middle Range Theorization of Energy Politics: The Struggle for Energy Efficient Appliances," *Environmental Politics* 20 (2011): 705–726.
46. Perrow, "Organizations and Global Warming."
47. Aaron McCright and Riley Dunlap, "Anti-reflexivity: The American Conservative Movement's Success in Undermining Climate Science and Policy," *Theory, Culture and Society* 27, no. 2–3 (2010): 100–133.
48. E.g., Arthur Mol, *Globalization and Environmental Reform* (Cambridge, MA: MIT Press, 2001); Kenneth Gould, David Pellow, and Allan Schnaiberg, *The Treadmill of Production: Injustice and Unsustainability in the Global Economy* (Boulder, CO: Paradigm, 2008).
49. Robert Antonio and Brett Clark, "The Climate Change Divide in Social Theory," in *Climate Change and Society: Sociological Perspectives*, ed. Riley Dunlap and Robert Brulle (New York: Oxford University Press, 2015), 333–368.
50. E.g., Bruce Greenwald and Joseph Stiglitz, "New and Old Keynesians," *Journal of Economic Perspectives* 7, no. 1 (1993): 23–44.
51. Freudenburg, "Privileged Access, Privileged Accounts."

52. E.g., Roberts and Parks, *A Climate of Injustice*. See also Andrew Jorgenson and Brett Clark, "Are the Economy and the Environment Decoupling?: A Comparative International Study, 1960–2005," *American Journal of Sociology* 118 (2012): 1–44; Richard York, Eugene Rosa, and Thomas Dietz, "Footprints on the Earth: The Environmental Consequences of Modernity," *American Sociological Review* 68 (2003): 279–300.
53. E.g., Robert Bullard, "Solid Waste Sites and the Black Houston Community," *Sociological Inquiry* 53 (1983): 273–288; Paul Mohai, David Pellow, and Timmons Roberts, "Environmental Justice," *Annual Review of Environmental Resources* 34 (2009): 404–430.
54. Don Grant, Albert Bergesen, and Andrew Jones, "Organizational Size and Pollution: The Case of the U.S. Chemical Industry," *American Sociological Review* 67 (2002): 389–407; Harland Prechel and Alesha Istvan, "Disproportionality of Corporations' Environmental Pollution in the Electrical Energy Industry," *Sociological Perspectives* 59, no. 3 (2016): 505–527.
55. E.g., Anya Robertson and Mary Collins, "Super Emitters in the United States Coal-Fired Electric Utility Industry: Comparing Disproportionate Emissions Across Facilities and Parent Companies," *Environmental Sociology* 5 (2019): 70–81.
56. George Kingsley Zipf, *Human Behavior and the Principle of Least Effort: An Introduction to Human Ecology* (Cambridge, MA: Addison-Wesley, 1949; Eastford, CT: Martino, 2012); Freudenburg, "Privileged Access, Privileged Accounts."
57. E.g., Richard H. Thaler, *Misbehaving: The Making of Behavioral Economics* (New York: Norton, 2015); D. Kahneman, *Thinking, Fast and Slow* (New York: Farrar, Straus and Giroux, 2011); Robert Shiller, *Finance and the Good Society* (Princeton, NJ: Princeton University Press, 2012).
58. E.g., Paula Castesana and Salvador Puliafito, "Development of an Agent-Based Model and Its Application to the Estimation of Global Carbon Emissions," *Low Carbon Economy* 4 (2014): 24–34.
59. Benjamin Sovacool, Jonn Axsen, and Steve Sorrell, "Promoting Novelty, Rigor, and Style in Energy Social Science: Towards Codes of Practice for Appropriate Methods and Research Design," *Energy Research and Social Science* 45 (2018): 12–42.
60. Gianluca Manzo, "Potentialities and Limitations of Agent-Based Simulations," *Review of French Sociology* 55 (2014): 653–688.

61. For a comprehensive review of these frameworks and their limitations from a social science perspective, see Sovacool, Axsen, and Sorrell, "Promoting Novelty, Rigor, and Style in Energy Social Science."
62. Mark Granovetter, "Economic Structure and Social Action: The Problem of Embeddedness," *American Journal of Sociology* 91, no. 3 (1985): 481–510; Peter Evans, *Embedded Autonomy: States and Industrial Transformation* (Princeton, NJ: Princeton University Press, 1995).
63. For an examination of the conditions under which corporations are likely to invest in energy efficiency technologies and other measures designed to address climate change, see Michael P. Vandenbergh and Jonathan M. Gilligan, *Beyond Politics: The Private Governance Response to Climate Change* (New York: Cambridge University Press, 2017).
64. For an in-depth discussion of rebounds or efficiency spillovers at the micro level, see Andreas Nilsson, Magnus Bergquist, and Wesley P. Schultz, "Spillover Effects in Environmental Behaviors, Across Time and Context: A Review and Research Agenda," *Environmental Education Research* 23, no. 4 (2017): 573–589, https://doi.org/10.1080/13504622.2016.1250148; Heather Barnes Truelove et al., "Positive and Negative Spillover of Pro-environmental Behavior: An Integrative Review and Theoretical Framework," *Global Environmental Change* 29 (2014): 127–138; Ellen Van Der Werff and Linda Steg, "Spillover Benefits: Emphasizing Different Benefits of Environmental Behavior and Its Effects on Spillover," *Frontiers in Psychology* 9 (2018), https://doi.org/10.3389/fpsyg.2018.02347.
65. William Stanley Jevons, *The Coal Question: An Inquiry Concerning the Progress of the Nation, and the Probable Exhaustion of Our Coal-Mines*, 3rd ed., ed. A. W. Flux (1865; New York: Augustus M. Kelley, 1965).
66. Jianguo Liu et al., "Complexity of Coupled Human and Natural Systems," *Science* 317 (2007): 1513–1516; Jianguo Liu et al., "Framing Sustainability in a Telecoupled World," *Ecology and Society* 18, no. 2 (2013): 26, http://dx.doi.org/10.5751/ES-05873-180226.
67. Max Weber, *The Protestant Ethic and the Spirit of Capitalism*, trans. T. Parsons (1904–1905; London: Routledge, 1992).
68. E.g., Richard York and Julius McGee, "Understanding the Jevons Paradox," *Environmental Sociology* 2 (2016): 77–87.
69. Michael Hannan and John Freeman, "Structural Inertia and Organizational Change," *American Sociological Review* 49, no. 2 (1984): 149–164.

70. Patricia Bromley and Walter Powell, "From Smoke and Mirrors to Walking the Talk: Decoupling in the Contemporary World," *Academy of Management Annals* 6 (2012): 483–530; Wade Cole, "Mind the Gap: State Capacity and the Implementation of Human Rights Treaties," *International Organization* 69 (2005): 405–441.
71. See, e.g., Thomas P. Lyon and John W. Maxwell, "Greenwash: Corporate Environmental Disclosure Under Threat of Audit," *Journal of Economics and Management Strategy* 20 (2011): 3–41.
72. James D. Westphal and Edward J. Zajac, "Decoupling Policy from Practice: The Case of Stock Repurchase Programs," *Administrative Science Quarterly* 46 (2001): 202–228.
73. Victor, *Global Warming Gridlock.*
74. Paul Hirsch, Stuart Michaels, and Ray Friedman, " 'Dirty Hands' Versus 'Clean Models': Is Sociology in Danger of Being Seduced by Economics?" *Theory and Society* 16, no. 3 (1987): 317–336.
75. Antony Millner and Thomas McDermott, "Model Confirmation in Climate Economics," *Proceedings of the National Academy of Sciences* 113, no. 31 (2016): 8675–8680.
76. R. Herendeen and J. Tanaka, "Energy Cost of Living," *Energy* 1 (1976): 165–178; Emily Huddart Kennedy, Harvey Krahn, and Naomi Krogman, "Egregious Emitters: Disproportionality in Household Carbon Footprints," *Environment and Behavior* 46, no. 5 (2014): 535–555.
77. Glen Peters et al., "Growth in Emission Transfers Via International Trade from 1990 to 2008," *Proceedings of the National Academy of Sciences* 108, no. 21 (2011): 8903–8908.
78. Richard Heede, "Tracing Anthropocentric Carbon Dioxide and Methane Emissions to Fossil Fuel and Cement Producers, 1854–2010," *Climatic Change* 122, no. 1–2 (2014): 229–241; Alison Kirsch, *Banking on Climate Change: Fossil Fuel Finance Report Card 2019* (San Francisco: Rainforest Action Network, 2019).
79. Monica Prasad, "Problem-Solving Sociology," *Contemporary Sociology* 47 (2018): 393–398.
80. Michael Burawoy, "For Public Sociology," *American Sociological Review* 70, no. 1 (2005): 4–28; Theda Skocpol, "How the Scholars Strategy Network Helps Academics Gain Public Influence," *Perspectives on Politics* 12, no. 3 (2014): 695–703.

81. McCright and Dunlap, "Anti-reflexivity"; Thomas Dietz, Rachael L. Shwom, and C. T. Whitley, "Climate Change and Society," *Annual Review of Sociology* (forthcoming).
82. Intergovernmental Panel on Climate Change (IPCC), *Global Warming of 1.5OC* (Geneva: IPCC, 2018), https://www.ipcc.ch/sr15/download/.
83. Dietz et al., "Household Actions Can Provide a Behavioral Wedge."
84. Vandenbergh and Gilligan, *Beyond Politics*.

2. CLEANING UP THEIR ACT: POTENTIAL EMISSION REDUCTIONS FROM TARGETING THE WORST OF THE WORST POWER PLANTS

1. Robert Kunzig, "The Will to Change," *National Geographic*, November 2015, 42.
2. Kerstine Appunn, Yannick Haas, and Julien Wettengel, "Germany's Energy and Consumption Power Mix in Charts," *Clean Energy Wire*, January 2020, https://www.cleanenergywire.org/factsheets/germanys-energy-consumption-and-power-mix-charts.
3. William Wilkes, Hayley Warren, and Brian Parkin, "Germany's Failed Climate Goals," *Bloomberg*, August 15, 2018, https://www.bloomberg.com/graphics/2018-germany-emissions/; Sören Amelang, "Germany to Miss Climate Targets 'Disastrously': Leaked Government Paper," *Climate Home News*, November 10, 2017, https://www.climatechangenews.com/2017/10/11/germany-miss-climate-targets-disastrously-leaked-government-paper/.
4. Another complicating factor is that in Germany the regions with the highest renewables potential are not the regions with the highest demand for electricity and getting large-scale transmission lines sited is very difficult because of permitting and local opposition.
5. A sector usually refers to an area of the economy in which businesses produce the same or similar products or services. It can also be thought of as an industry or market that shares common operating characteristics, such as the electricity generation sector. International Energy Agency (IEA), *Energy Technology Perspectives* (Paris: OECD/IEA, 2012); G. Marland, T. A. Boden, and R. J. Andres, "Global, Regional, and National Fossil Fuel CO_2 Emissions," in "Trends: A Compendium of Data on Global Change," Carbon Dioxide Information Analysis Center, Oak Ridge

National Laboratory, U.S. Department of Energy, Oak Ridge, TN, 2008 https://cdiac.ess-dive.lbl.gov/trends/emis/overview.html.

6. International Energy Agency (IEA), *World Energy Outlook* (Paris: IEA, 2009).
7. Intergovernmental Panel on Climate Change (IPCC), *Climate Change 2014: Synthesis Report* (Geneva: IPCC, 2014).
8. International Energy Agency, *Energy Technology Perspectives*.
9. International Energy Agency (IEA), *Sectoral Approaches in Electricity: Building Bridges to a Safe Climate* (Paris: IEA, 2009); International Energy Agency (IEA), *How the Energy Sector Can Deliver on a Climate Agreement in Copenhagen* (Paris: IEA, 2009); Center for Clean Air Policy (CCAP), *Sectoral Approaches: A Pathway to Nationally Appropriate Mitigation Actions* (Washington, DC: CCAP, 2008), http://ccap.org/assets/Center-for-Clean-Air-Policy-Interim-Report-Sectoral-Approaches-A-Pathway-to-Nationally-Appropriate-Mitigation-Actions_CCAP-December-20081.pdf; Christian Egenhofer and Noriko Fujiwara, *Global Sectoral Industry Approaches to Global Climate Change: The Way Forward* (Brussels: Centre for European Policy Studies Task Force, 2008), https://www.ceps.eu/ceps-publications/global-sectoral-industry-approaches-climate-change-way-forward/.
10. International Energy Agency, *Sectoral Approaches in Electricity*; World Resources Institute (WRI), *Target: Intensity, an Analysis of Greenhouse Gas Intensity Targets* (Washington, DC: WRI, 2006).
11. Andrew Jorgenson and Brett Clark, "Are the Economy and the Environment Decoupling?: A Comparative International Study, 1960–2005," *American Journal of Sociology* 118 (2012): 1–44; Richard York et al., "It's a Material World: Trends in Material Extraction in China, India, Indonesia, and Japan," *Nature and Culture* 6, no. 2 (2011): 103–122.
12. Also, while the sectoral approach makes sense for engineers, it makes less sense for sociologists. For example, it splits transportation and residential, so that household decisions about commuting are lumped with corporate decisions about fleet composition, and it minimizes the impacts of household decisions because in-home energy use is in a different category than household transportation energy use even though the same family is making these decisions. For these reasons, some sociologists, in collaboration with scholars from other disciplines, have focused on studying behavioral wedges and have called for analyses that

are based on who makes decisions, not on how engineers lump technologies together, and that emphasize the supply chain. Thomas Dietz et al., "Household Actions Can Provide a Behavioral Wedge to Rapidly Reduce U.S. Carbon Emissions," *Proceedings of the National Academy of Sciences* 106 (2009): 18452–18456.

13. World Resources Institute, *Target*.
14. William Freudenburg, "Privileged Access, Privileged Accounts: Toward a Socially Structured Theory of Resources and Discourses," *Social Forces* 84 (2005): 89–114; William Freudenburg, "Environmental Degradation, Disproportionality, and the Double Diversion: Reaching Out, Reaching Ahead, and Reaching Beyond," *Rural Sociology* 71, no. 1 (2006): 3–32.
15. The Gini coefficient is most often used in social science research on income and wealth inequities. It measures the area between the Lorenz curve and the hypothetical line of absolute equality, expressed as a proportion of the maximum area under the line.
16. Freudenburg, "Privileged Access, Privileged Accounts."
17. Anya Robertson and Mary Collins, "Super Emitters in the United States Coal-Fired Electric Utility Industry: Comparing Disproportionate Emissions Across Facilities and Parent Companies," *Environmental Sociology* 5 (2019): 70–81.
18. Freudenberg's approach to analyzing disproportionality in the production of pollution and emissions has also been used in studies that link industrial pollution disproportionalities to public health disparities and environmental justice communities within the United States as well as disproportionality in pollution among industrial parent companies and to disproportionality in household carbon footprints within one city. Mary Collins, "Risk-Based Targeting: Identifying Disproportionalities in the Sources and Effects of Industrial Pollution," *American Journal of Public Health* 101 (2011): 231–237; Mary Collins, Ian Munoz, and Joseph JaJa, "Linking 'Toxic Outliers' to Environmental Justice Communities," *Environmental Research Letters* 11 (2016): 015004; Harland Prechel and Alesha Istvan, "Disproportionality of Corporations' Environmental Pollution in the Electrical Energy Industry," *Sociological Perspectives* 59, no. 3 (2016): 505–527; Emily Huddart Kennedy, Harvey Krahn, and Naomi Krogman, "Egregious Emitters: Disproportionality in Household Carbon Footprints," *Environment and Behavior* 46, no. 5 (2014): 535–555.

19. David Wheeler and Kevin Ummel, "Calculating CARMA: Global Estimation of CO_2 Emissions from the Power Sector," Working Paper 145, Center for Global Development, Washington, DC, https://www.cgdev.org/publication/calculating-carma-global-estimation-co2-emissions-power-sector-working-paper-145.
20. Thomas Dietz et al., "Household Actions Can Provide a Behavioral Wedge."
21. As a reminder, 2009 is the most recent year for which these data are currently available.
22. In Stata, we used the "egen_inequal" module, which provides a series of programs for calculating various inequality measures, including the Gini coefficient, and also allows for the weighting of cases by a specified variable in the calculation of Gini coefficients.
23. The emissions data were obtained from the Center for Global Development's Carbon Monitoring for Action (CARMA) database. David Wheeler and Kevin Ummel, "Calculating CARMA: Global Estimation of CO_2 Emissions from the Power Sector" (Working Paper 145, Center for Global Development, Washington, DC, 2008), https://www.cgdev.org/publication/calculating-carma-global-estimation-co2-emissions-power-sector-working-paper-145.
24. For the 19,941 power plants in the data set, the Pearson's correlation coefficient for plant-level output and plant size (measured as megawatt capacity) is .90 (.001 level of statistical significance, two-tailed test), which suggests inefficiencies among some of the larger power plants and emphasizes the importance of weighting plants by output instead of by size when calculating the disproportionality Gini coefficients.
25. The Pearson's correlation coefficient is commonly used to measure the strength of the association between two variables. It can range in value from –1 to +1, where a positive value that is close to +1 indicates a strong positive relationship between the two variables and a negative value that is close to –1 indicates a strong negative relationship. A value of 0 suggests no association at all.
26. Andrew Jorgenson et al., "Social Science Perspectives on Drivers of and Responses to Global Climate Change," *Wiley Interdisciplinary Reviews: Climate Change* 10, no. 1 (2019): e554, https://doi.org/10.1002/wcc.554.

27. Thomas Dietz, Scott Frey, and Linda Kalof, "Estimation with Cross-National Data: Robust and Resampling Estimators," *American Sociological Review* 52 (1987): 380–390.
28. Andrew Jorgenson, "Economic Development and the Carbon Intensity of Human Well-Being," *Nature Climate Change* 4 (2014): 186–189; Jorgenson and Clark, "Are the Economy and the Environment Decoupling?"; Eugene Rosa and Thomas Dietz, "Human Drivers of National Greenhouse Gas Emissions," *Nature Climate Change* 2 (2012): 581–586; Richard York, Eugene Rosa, and Thomas Dietz, "Footprints on the Earth: The Environmental Consequences of Modernity," *American Sociological Review* 68 (2003): 279–300.
29. The variance inflation factors (VIFs) have a mean of 1.40 and a range of 1.20–1.52 in model 1, a mean of 2.62 and a range of 1.10–4.51 in model 2, and a mean of 1.46 and a range of 1.27–1.72 in model 3, indicating that the regression analyses are not affected by multicollinearity.
30. Jorgenson and Clark, "Are the Economy and the Environment Decoupling?"; Rosa and Dietz, "Human Drivers of National Greenhouse Gas Emissions"; Wesley Longhofer and Andrew Jorgenson, "Decoupling Reconsidered: Does World Society Integration Influence the Relationship Between the Environment and Economic Development?" *Social Science Research* 65 (2017): 17–29; Ryan Thombs, "The Transnational Tilt of the Treadmill and the Role of Trade Openness on Carbon Emissions: A Comparative International Study, 1965–2010," *Sociological Forum* 33 (2018): 422–442; York, Rosa, and Dietz, "Footprints on the Earth. "The Environmental Consequences of Modernity," *American Sociological Review* 68 (2003): 279-300.."
31. Morgan Energy Solutions, "Taichung Power Plant: World's Largest Coal Fired Plant," January 16, 2003.
32. "Russian Firm Studying World's Largest Coal-Fired Plant to Supply China," Reuters, May 26, 2014, https://www.reuters.com/article/russia-interrao-plant/russian-firm-studying-worlds-largest-coal-fired-plant-to-supply-china-idUSL6N0OC30R20140526.
33. See "List of Countries by Carbon Dioxide Emissions," 2019, http://en.wikipedia.org/wiki/List_of_countries_by_carbon_dioxide_emissions.
34. "Behemoth Coal Plants Threaten Utilities' CO2 Goals," Climatewire, August 23, 2019. https://www.eenews.net/stories/1061039369.

3. RECIPES FOR DISASTER: HOW SOCIAL STRUCTURES INTERACT TO MAKE ENVIRONMENTALLY DESTRUCTIVE PLANTS EVEN MORE SO

1. Riley Dunlap and Robert J. Brulle, eds., *Climate Change and Society: Sociological Perspectives* (New York: Oxford University Press, 2015); Eugene Rosa and Thomas Dietz, "Human Drivers of National Greenhouse Gas Emissions," *Nature Climate Change* 2 (2012): 581–586.
2. Thomas Burns, Byron Davis, and Edward Kick, "Position in the World-System and National Emissions of Greenhouse Gases," *Journal of World-Systems Research* 3, no. 3 (1997): 432–466; Bruce Podobnik, *Global Energy Shifts: Fostering Sustainability in a Turbulent Age* (Philadelphia: Temple University Press, 2006); Timmons Roberts and Bradley Parks, *A Climate of Injustice: Global Inequality, North-South Politics, and Climate Policy* (Cambridge, MA: MIT Press, 2007).
3. Evan Schofer and Ann Hironaka, "The Effects of World Society on Environmental Protection Outcomes," *Social Forces* 84 (2005): 25–47; Ann Hironaka, *Greening the Globe: World Society and Environmental Change* (New York: Cambridge University Press, 2014); Jennifer Givens, "World Society, World Polity, and the Carbon Intensity of Well-Being, 1990–2011," *Sociology of Development* 3 (2017): 403–435.
4. Rachael Shwom, "Strengthening Sociological Perspectives on Organizations and the Environment," *Organization and Environment* 22 (2009): 271–292; Mario E. Bergara, Witold J. Henisz, and Pablo T. Spiller, "Political Institutions and Electric Utility Investment: A Cross-Nation Analysis," *California Management Review* 40 (1998): 18–35; Harland Prechel and Lu Zheng, "Corporate Characteristics, Political Embeddedness, and Environmental Pollution by Largest U.S. Corporations," *Social Forces* 90 (2012): 947–970; Thomas K. Rudel, "How Do People Transform Landscapes?: A Sociological Perspective on Suburban Sprawl and Tropical Deforestation," *American Journal of Sociology* 115 (2009): 129–154.
5. Charles Perrow, "Organizations and Global Warming," in *Routledge Handbook of Climate Change and Society*, ed. Constance Lever-Tracy (New York: Routledge, 2010), 59–77; Charles Perrow and Simone Pulver, "Organizations and Markets," in *Climate Change and Society:*

Sociological Perspectives, ed. R. Dunlap and R. Brulle (New York: Oxford University Press, 2005).

6. In this study, structure is understood as the relatively durable features of an organization and the larger systems and institutions in which it operates that enhance or constrain its access to resources and define the rules governing legitimate action.
7. In this study, interaction is understood in two senses. The first, which is the more conventional one, is associated with quantitative, variable-based approaches like regression. It is premised on assumptions about sampling and the distribution of data and refers to the extent to which the average effect of two factors together exceeds the average effect of each factor considered individually. The second is associated with qualitative, case-oriented approaches like qualitative comparative analysis (QCA) and its fuzzy-set version (fsQCA). It is based on an examination of sets, subsets, and supersets of cases and refers to the combinations of conditions that are necessary and sufficient to produce an outcome. We use the latter approach to identify the sets of global, political, and organizational conditions associated with power plants' high CO_2 emission rates.
8. Intergovernmental Panel on Climate Change (IPCC), *Climate Change 2014: Synthesis Report* (Geneva: IPCC, 2014).
9. Richard York, Eugene Rosa, and Thomas Dietz, "Footprints on the Earth: The Environmental Consequences of Modernity," *American Sociological Review* 68 (2003): 279–300; Thomas Dietz, Eugene Rosa, and Richard York, "Human Driving Forces of Global Change: Dominant Perspectives," in *Human Footprints on the Global Environment: Threats to Sustainability*, ed. E. Rosa et al. (Cambridge, MA: MIT Press, 2010), 83–134; Thomas Dietz, "Prolegomenon to a Structural Human Ecology of Human Well-Being," *Sociology of Development* 1 (2015): 123–148.
10. Andrew Jorgenson and Brett Clark, "Are the Economy and the Environment Decoupling?: A Comparative International Study, 1960–2005," *American Journal of Sociology* 118 (2012): 1–44; Ryan Thombs and Xiaorui Huang, "Uneven Decoupling: The Economic Growth–CO_2 Emissions Relationship in the Global North, 1870–2014," *Sociology of Development* 5 (2019): 410–427.
11. Rachael Shwom, "A Middle Range Theorization of Energy Politics: The Struggle for Energy Efficient Appliances," *Environmental Politics* 20 (2011): 705–726.

12. Burns, Davis, and Kick, "Position in the World-System"; Andrew Jorgenson, Christopher Dick, and Matthew C. Mahutga, "Foreign Investment Dependence and the Environment: An Ecostructural Approach," *Social Problems* 54 (2007): 371–394; Roberts and Parks, *A Climate of Injustice*.
13. Christopher Chase-Dunn and Peter Grimes, "World-Systems Analysis," *Annual Review of Sociology* 21 (1995): 387–417.
14. Podobnik, *Global Energy Shifts*.
15. Jorgenson, Dick, and Mahutga, "Foreign Investment Dependence and the Environment; Roberts and Parks, *A Climate of Injustice*; For similar arguments within environmental sociology, see the metabolic rift and treadmill of production frameworks, which suggest there is an incompatibility and "enduring conflict" between economic production and ecological limits: John Bellamy Foster, Brett Clark, and Richard York, *The Ecological Rift: Capitalism's War on the Earth* (New York: Monthly Review Press, 2010); Kenneth Gould, David Pellow, and Allan Schnaiberg, *The Treadmill of Production: Injustice and Unsustainability in the Global Economy*, (Boulder, CO: Paradigm, 2008).
16. Wesley Longhofer and Evan Schofer, "National and Global Origins of Environmental Association," *American Sociological Review* 75 (2010): 505–533; Schofer and Hironaka, "The Effects of World Society."
17. David John Frank, Ann Hironaka, and Evan Schofer, "The Nation-State and the Natural Environment Over the Twentieth Century," *American Sociological Review* 65 (2000): 96–116; Hironaka, *Greening the Globe*; Schofer and Hironaka, "The Effects of World Society."
18. Hironaka, *Greening the Globe*; Schofer and Hironaka, "The Effects of World Society"; John M. Shandra et al., "Ecologically Unequal Exchange, World Polity, and Biodiversity Loss: A Cross-National Analysis of Threatened Mammals," *International Journal of Comparative Sociology* 50 (2009): 285–310; Jennifer Givens and Andrew Jorgenson, "Individual Environmental Concern in the World Polity," *Social Science Research* 42 (2013): 418–431. For similar arguments within environmental sociology, see the ecological modernization perspective (which argues, contrary to the treadmill of production and metabolic rift perspectives, that capitalism is compatible with environmental protection). Arthur P. J. Mol, Gert Spaargaren, and David A. Sonnenfeld, "Ecological Modernization Theory: Taking Stock, Moving Forward," in *The Routledge International Handbook of Social and Environmental*

Change, ed. Stewart Lockie, David Sonnenfeld, and Dana R. Fisher (New York: Routledge, 2014), 15–30; Dana R. Fisher and William R. Freudenburg, "Post Industrialization and Environmental Quality: An Empirical Analysis of the Environmental State," *Social Forces* 83 (2004): 157–188.

19. Frederick H. Buttel, "The Treadmill of Production: An Appreciation, Assessment, and Agenda for Research," *Organization and Environment* 17 (2004): 323–336.
20. Prechel and Zheng, "Corporate Characteristics"; Rudel, "How Do People Transform Landscapes?"
21. Shwom, "Strengthening Sociological Perspectives."
22. Harland Prechel and Alesha Istvan, "Disproportionality of Corporations' Environmental Pollution in the Electrical Energy Industry," *Sociological Perspectives* 59, no. 3 (2016): 505–527. See also Thomas Dietz et al., "Political Influences on Greenhouse Gas Emissions from US States," *Proceedings of the National Academy of Sciences* 112 (2015): 8254–8259.
23. Bergara, Henisz, and Spiller, "Political Institutions and Electric Utility Investment." See also Witold J. Henisz, "The Institutional Environment for Infrastructure Investment," *Industrial and Corporate Change* 11 (2002): 355–389; Witold J. Henisz, Bennet A. Zelner, and Mauro F. Guillén, "The Worldwide Diffusion of Market-Oriented Infrastructure Reform, 1977–1999," *American Sociological Review* 70 (2005): 871–897.
24. United Nations Framework Convention on Climate Change, "Bali Action Plan," in "Report of the Conference of the Parties on Its Thirteenth Session," Bonn, 2007, https://unfccc.int/resource/docs/2007/cop13/eng/06a01.pdf.
25. Perrow, "Organizations and Global Warming."
26. W. Richard Scott, *Organizations: Rational, Natural and Open Systems*, 3rd ed. (Englewood Cliffs, NJ: Prentice-Hall, 1992), 258–267.
27. See also Don Grant, Albert Bergesen, and Andrew Jones, "Organizational Size and Pollution: The Case of the U.S. Chemical Industry," *American Sociological Review* 67 (2002): 389–407.
28. In addition to power relations, there may be diseconomies of organizational scale. Thomas S. Lough, "Energy Analysis of the Structures of Industrial Organizations," *Energy* 21, no. 2 (1996): 131–139.
29. Prechel and Istvan, "Disproportionality of Corporations' Environmental Pollution"; Harland Prechel and George Touche, "The Effects of

Organizational Characteristics and State Environmental Policies on Sulfur-Dioxide Pollution in U.S. Electrical Energy Corporations," *Social Science Quarterly* 95 (2014): 76–96.

30. Michael T. Hannan and John Freeman, "Structural Inertia and Organizational Change," *American Sociological Review* 49, no. 2 (1984): 149–164.
31. Perrow, "Organizations and Global Warming."
32. Peer C. Fiss, "A Set-Theoretic Approach to Organizational Configurations," *Academy of Management Review* 32 (2007): 1180–1198; Peer C. Fiss, Bart Cambré, and Axel Marx, eds., *Configurational Theory and Methods in Organizational Research* (Bingley, UK: Emerald Group, 2013).
33. Donal Crilly, "Corporate Social Responsibility: A Multilevel Explanation of Why Managers Do Good," *Research in Sociology of Organizations* 58 (2013): 190–213.
34. Andrew Jorgenson, Christopher Dick, and John M. Shandra, "World Economy, World Society, and Environmental Harms in Less-Developed Countries," *Sociological Inquiry* 81 (2011): 53–87; Kristen Shorette, "Outcomes of Global Environmentalism: Longitudinal and Cross-National Trends in Chemical Fertilizer and Pesticide Use," *Social Forces* 91 (2012): 299–325; Frederick H. Buttel, "World Society, the Nation-State, and Environmental Protection: Comment on Frank, Hironaka, and Schofer," *American Sociological Review* 65 (2000): 117–121.
35. Perrow and Pulver, "Organizations and Markets"; Christopher Marquis, Michael W. Toffel, and Yanhua Zhou, "Scrutiny, Norms, and Selective Disclosure: A Global Study of Greenwashing," *Organization Science* 27 (2016): 483–504.
36. Robert Bullard, "Solid Waste Sites and the Black Houston Community," *Sociological Inquiry* 53 (1983): 273–288; David Schelly and Paul Stretesky, "An Analysis of the 'Path of Least Resistance' Argument in Three Environmental Justice Success Cases," *Society and Natural Resources* 22 (2009): 369–380.
37. Kerry Ard, "Trends in Exposure to Industrial Air Toxins for Different Racial and Socioeconomic Groups: A Spatial and Temporal Examination of Environmental Inequality in the U.S. from 1995 to 2004," *Social Science Research* 53 (2015): 375–390; Robert Brulle and David Pellow, "Environmental Justice: Human Health and Environmental Inequalities," *Annual Review of Public Health* 27 (2006): 103–124; Andrew Jorgenson et al., "Power, Proximity, and Physiology: Does Income

Inequality and Racial Composition Amplify the Impacts of Air Pollution on Life Expectancy in the United States?" *Environmental Research Letters* (2020), https://doi.org/10.1088/1748-9326/ab6789.

38. Rob Clark and Jason Beckfield, "A New Trichotomous Measure of World-System Position Using the International Trade Network," *International Journal of Comparative Sociology* 50 (2009): 5–38.
39. Max Weber, *Economy and Society: An Outline of Interpretive Sociology* (1922; Berkeley: University of California Press, 1978).
40. Thomas E. Shriver, Alison E. Adams, and Chris M. Messer, "Power, Quiescence, and Pollution: The Suppression of Environmental Grievances," *Social Currents* 1 (2014): 275–292.
41. Weber, *Economy and Society.*
42. Bergara, Henisz, and Spiller, "Political Institutions and Electric Utility Investment." See also Henisz, "The Institutional Environment for Infrastructure Investment."
43. Chase-Dunn and Grimes, "World-Systems Analysis."
44. Hannan and Freeman, "Structural Inertia and Organizational Change."
45. The raw and unique coverage scores indicate, respectively, which share of the outcome is explained by a certain alternative path and which share of the outcome is exclusively explained by a certain alternative path. Table 3.1 reveals that each pathway or configuration explains 43.9 to 52.5 percent of high CO_2 emission rates and exclusively explains 1 to 5 percent of this outcome. These configurations' solution consistency scores (0.880 to 0.960) suggest that they are very nearly sufficient to produce a high rate of CO_2 emissions. Finally, the coverage and consistency scores for the five configurations combined (0.739 and 0.830) suggest the total solution explains a large number of cases and often leads to high emission rates.
46. The BIC statistic measures the relative likelihood of the models given the data. Models with more negative BIC scores have a higher relative likelihood.
47. Claude Rubinson, "Presenting Qualitative Comparative Analysis: Notation, Tabular Layout, and Visualization," *Methodological Innovations* 12 (May–August 2019): 1–22.
48. Clark and Beckfield, "A New Trichotomous Measure."
49. Dunlap and Brulle, *Climate Change and Society*; Joane Nagel, Jeffrey Broadbent, and Thomas Dietz, *Workshop on Sociological Perspectives on Climate Change* (Washington, DC: American Sociological Association, 2010).

50. York, Rosa, and Dietz, "Footprints on the Earth."
51. United Nations Framework Convention on Climate Change, "Bali Action Plan"; International Energy Agency (IEA), *Sectoral Approaches in Electricity: Building Bridges to a Safe Climate* (Paris: IEA, 2009).
52. Chase-Dunn and Grimes, "World-Systems Analysis."
53. Hironaka, *Greening the Globe*. See also Rudel, "How Do People Transform Landscapes?"
54. Neil Fligstein and Doug McAdam, *A Theory of Fields* (New York: Oxford University Press, 2015).
55. Doug McAdam and Hilary Boudet, *Putting Social Movements in Their Place: Opposition to Energy Projects, 2000–2005* (New York: Cambridge University Press, 2012).
56. Scott, *Organizations*.
57. Perrow, "Organizations and Global Warming"; Perrow and Pulver, "Organizations and Markets."
58. Hannan and Freeman, "Structural Inertia and Organizational Change."
59. Paul J. DiMaggio and Walter W. Powell, "The Iron Cage Revisited: Institutional Isomorphism and Collective Rationality in Organizational Fields," *American Sociological Review* 48 (1983): 147–160.
60. Danny Miller, "The Genesis of Configuration," *Academy of Management Review* 12 (1987): 686–701.
61. Peter L. Berger and Thomas Luckman, *The Social Construction of Reality* (New York: Doubleday, 1967).
62. Andrew Jorgenson et al., "Social Science Perspectives on Drivers of and Responses to Global Climate Change." *Wiley Interdisciplinary Reviews: Climate Change* 10, no. 1 (2019): e554, https://doi.org/10.1002/wcc.554; Simone Pulver, "Corporate Responses," in *The Oxford Handbook of Climate Change and Society*, ed. J. Dryzek, R. Norgaard, and D. Schlosberg (Oxford: Oxford University Press, 2011), 581–592; Rosa and Dietz, "Human Drivers."
63. Paul McLaughlin, "Climate Change, Adaptation, and Vulnerability: Reconceptualizing Societal-Environment Interaction Within a Socially Constructed Adaptive Landscape," *Organization and Environment* 24 (2011): 269–291. See also Thomas E. Shriver, Alison E. Adams, and Sherry Cable, "Discursive Obstruction and Elite Opposition to Environmental Activism in the Czech Republic," *Social Forces* 91 (2013): 873–893; Justin Farrell, "Network Structure and Influence of the

Climate Change Counter-Movement," *Nature Climate Change* 6, no. 4 (2016): 370–374.

4. A WIN-WIN SOLUTION?: THE PARADOXICAL EFFECTS OF EFFICIENCY ON PLANTS' CO_2 EMISSIONS

1. William Stanley Jevons, *The Coal Question: An Inquiry Concerning the Progress of the Nation, and the Probable Exhaustion of Our Coal-Mines*, 3rd ed., ed. A. W. Flux (1865; New York: Augustus M. Kelley, 1965).
2. Timmons Roberts, Peter Grimes, and Jodie Manale, "Social Roots of Global Environmental Change: A World-Systems Analysis of Carbon Dioxide Emissions," *Journal of World-Systems Research* 9, no. 2 (2003): 277–315.
3. Kenneth Gould, David Pellow, and Allan Schnaiberg, *The Treadmill of Production: Injustice and Unsustainability in the Global Economy* (Boulder, CO: Paradigm, 2008).
4. Richard York, "Ecological Paradoxes: William Stanley Jevons and the Paperless Office," *Human Ecology Review* 13, no. 2 (2006): 143–147.
5. T. E. Graedel and Braden Allensby, *Industrial Ecology*, 2nd ed. (Englewood Cliffs, NJ: Prentice Hall, 2003).
6. Arthur P. J. Mol, Gert Spaargaren, and David A. Sonnenfeld, "Ecological Modernization Theory: Taking Stock, Moving Forward," in *The Routledge International Handbook of Social and Environmental Change*, ed. Stewart Lockie, David Sonnenfeld, and Dana R. Fisher (New York: Routledge, 2014), 15–30.
7. T. A. Cavlovic et al., "A Meta-analysis of Environmental Kuznets Curve Studies," *Journal of Industrial Ecology* 4, no. 4 (2000): 13–29.
8. Intergovernmental Panel on Climate Change (IPCC), *Climate Change 2014: Synthesis Report* (Geneva: IPCC, 2014), https://www.ipcc.ch/report/ar5/wg2/.
9. Richard Campbell, *Increasing the Efficiency of Existing Coal-Fired Power Plants* (Washington, DC: Congressional Research Service, 2013), https://fas.org/sgp/crs/misc/R43343.pdf; International Energy Agency (IEA), *Upgrading and Efficiency Improvements in Coal-Fired Power Plants* (Paris: IEA, 2010), http://iea-coal.org/uk/site/2010/publications-section/reports; Asia-Pacific Economic Cooperation (APEC), Energy

Working Group, *Costs and Effectiveness of Upgrading and Refurbishing Older Coal-Fired Power Plants in Developing APEC Economies* (Singapore: APEC, 2005), www.egcfe.ewg.apec.org/projects/UpgradePP_Report_2005.pdf.

10. National Energy Technology Laboratory, *Reducing CO_2 Emissions by Improving the Efficiency of the Existing Coal-Fired Power Plant Fleet* (2008), www.netl.doe.gov/energy-analyses/pubs/CFPP%20Efficiency-FINAL.pdf; U.S. Environmental Protection Agency (EPA), *Energy Efficiency as a Low-Cost Resource for Achieving Carbon Emissions Reductions* (Washington, DC: EPA, 2009).
11. National Renewable Energy Laboratory (NREL), *Evaluating Renewable Portfolio Standards and Carbon Cap Scenarios in the U.S. Electric Sector*, NREL/TP-6A2-48258 (Golden, CO: NREL, 2010); McKinsey and Company, *Unlocking Energy Efficiency in the U.S. Economy* (New York: McKinsey, 2009), https://www.mckinsey.com/~/media/mckinsey/dotcom/client_service/epng/pdfs/unlocking%20energy%20efficiency/us_energy_efficiency_exc_summary.ashx.
12. L. Brookes, "A Low-Energy Strategy for the UK by G. Leach et al.: A Review and Reply," *Atom* 269 (1979): 3–8; J. Daniel Khazzoom, "Economic Implications of Mandated Efficiency Standards for Household Appliances," *Energy Journal* 1, no. 4 (1980): 85–89; J. Daniel Khazzoom, "Response to Besen and Johnson's Comment on Economic Implications of Mandated Efficiency Standards for Household Appliances," *Energy Journal* 3, no. 1 (1982): 117–124; Jevons, *The Coal Question*.
13. Janos Beér, "High Efficiency Electric Power Generation: The Environmental Role," *Progress in Energy Combustion Science* 33, no. 2 (2006): 107–134.
14. T. J. Garrett, "No Way Out? The Double-Bind in Seeking Global Prosperity Alongside Mitigated Climate Change," *Earth System Dynamics* 3 (2012): 1–17; Richard York, "The Paradox at the Heart of Modernity: The Carbon Efficiency of the Global Economy," *International Journal of Sociology* 40, no. 2 (2010): 6–22.
15. Steve Sorrell, "Jevons' Paradox Revisited: The Evidence for Backfire from Improved Energy Efficiency," *Energy Policy* 37 (2009): 1456–1469; Lorna Greening, David Greene, and Carmen Difiglio, "Energy Efficiency and Consumption—The Rebound Effect—A Survey," *Energy Policy* 28, no. 6 (2000): 389–401.

16. Paul Hirsch, Stuart Michaels, and Ray Friedman, " 'Dirty Hands' Versus 'Clean Models': Is Sociology in Danger of Being Seduced by Economics?" *Theory and Society* 16, no. 3 (1987): 317–336.
17. Amelia Keyes et al., "The Affordable Clean Energy Rule and the Impact of Emissions Rebound on Carbon Dioxide and Criteria Air Pollutant Emissions," *Environmental Research Letters* 14 (2019): 044018.
18. Richard R. Nelson and Sidney G. Winter, *An Evolutionary Theory of Economic Change* (Cambridge, MA: Belknap Press, 1982).
19. Michael T. Hannan and John Freeman, "Structural Inertia and Organizational Change," *American Sociological Review* 49, no. 2 (1984): 149–164.
20. Christopher Chase-Dunn and Peter Grimes, "World-Systems Analysis," *Annual Review of Sociology* 21 (1995): 387–417.
21. Brett Clark and Richard York, "Carbon Metabolism: Global Capitalism, Climate Change, and the Biospheric Rift," *Theory and Society* 34, no. 4 (2005): 391–428.
22. Andrew Jorgenson, Brett Clark, and Jeffrey Kentor, "Militarization and the Environment: A Panel Study of Carbon Dioxide Emissions and the Ecological Footprints of Nations, 1970–2000," *Global Environmental Politics* 10 (2010): 7–29.
23. Ann Hironaka, *Greening the Globe: World Society and Environmental Change* (New York: Cambridge University Press, 2004); Wesley Longhofer and Andrew Jorgenson, "Decoupling Reconsidered: Does World Society Integration Influence the Relationship Between the Environment and Economic Development?" *Social Science Research* 65 (2017): 17–29.
24. Patricia Bromley and Walter Powell, "From Smoke and Mirrors to Walking the Talk: Decoupling in the Contemporary World," *Academy of Management Annals* 6 (2012): 483–530.
25. Dana Fisher, "COP-15 in Copenhagen: How the Merging of Movements Left Civil Society Out in the Cold," *Global Environmental Politics* 10, no. 2 (2010): 11–17.
26. Sarah Adair, David Hoppock, and Jonas Monast, "New Source Review and Coal Plant Efficiency Gains: How New and Forthcoming Air Regulations Affect Outcomes," *Energy Policy* 70 (2014): 183–192.

5. BOTTOM-UP STRATEGIES: THE EFFECTIVENESS OF LOCAL POLICIES AND ACTIVISM

1. Juliet Eilpern, Brady Dennis, and Chris Mooney, "Trump Administration Sees a 7-Degree Rise in Global Temperatures by 2100," *Washington Post*, September 28, 2018.
2. Sanya Carley, "The Era of State Energy Policy Innovation: A Review of Policy Instruments," *Review of Policy Research* 28 (2011): 265–294.
3. Monica Prasad and Steven Munch, "State-Level Renewable Electricity Policies and Reductions in Carbon Emissions," *Energy Policy* 45 (2012): 237–242; Barry Rabe, "Racing to the Top, the Bottom, or the Middle of the Pack? The Evolving State Government Role in Environmental Protection," in *Environmental Policy: New Directions for the Twenty-First Century*, ed. N. Vig and M. Kraft, 30–53 (Washington, DC: CQ Press, 2010).
4. International Energy Agency (IEA), *Sectoral Approaches in Electricity: Building Bridges to a Safe Climate*, (Paris: IEA, 2009).
5. Evan Schofer and Ann Hironaka, "The Effects of World Society on Environmental Protection Outcomes," *Social Forces* 84 (2005): 25–47.
6. Azis Choudry and D. Kapoor, eds., *NGOization: Complicity, Contradictions and Prospects* (London: Zed Books, 2013).
7. Frederick H. Buttel, "World Society, the Nation-State, and Environmental Protection: Comment on Frank, Hironaka, and Schofer," *American Sociological Review* 65 (2000): 117–121.
8. Frederick Engels, *The Condition of the Working Class in England in 1844* (London: Sonnenschein, 1892).
9. David N. Pellow, "Environmental Inequality Formation: Toward a Theory of Environmental Justice," *American Behavioral Scientist* 43 (2000): 581–601.
10. W. Richard Scott and Gerald Davis, *Organizations and Organizing: Rational, Natural, and Open Perspectives* (New York: Routledge, 2016).
11. For a review of social movements and their consequences, see Marco Giugni, "Was It Worth the Effort?: The Outcomes and Consequences of Social Movements," *Annual Review of Sociology* 24 (1998): 371–393.
12. James Bushnell, Carla Peterman, and Catherine Wolfram, "Local Solutions to Global Problems: Climate Change Policies and Regulatory Jurisdiction," *Review of Environmental Economics and Policy* 2, no. 2 (2008): 175–193; H. Doremus and M. Hanemann, "Babies and Bathwater: Why the Clean Air Act's Cooperative Federalism Framework

Is Useful for Addressing Global Warming," *Arizona Law Review* 50 (2008): 799–834; Barry Rabe, "States on Steroids: The Intergovernmental Odyssey of American Climate Policy," *Review of Policy Research* 25 (2008): 105–128; Pew Center on Global Climate Change, *Climate Change 101: State Action* (Arlington, VA: Pew, 2006).

13. Rabe, "Racing to the Top."
14. Robert Michaels, "National Renewable Portfolio Standard: Smart Policy or Misguided Gesture?" *Energy Law Journal* 29 (2008): 79–119.
15. David G. Victor, Joshua C. House, and Sarah Joy, "A Madisonian Approach to Climate Policy," *Science* 309 (2005): 1820–1821.
16. Nicholas Lutsey and Daniel Sperling, "America's Bottom-Up Climate Change Mitigation Strategy," *Energy Policy* 36 (2008): 673–685; Thomas P. Lyon and Haitao Yin, "Why Do States Adopt Renewable Portfolio Standards?: An Empirical Investigation," *Energy Journal* 31, no. 3 (2010): 133–157; Steffen Jenner et al., "What Drives States to Support Renewable Energy?" *Energy Journal* 33 (2012): 1–12; Daniel Matisoff, "The Adoption of State Climate Change Policies and Renewable Portfolio Standards: Regional Diffusion or Internal Determinants?" *Review of Policy Research* 25 (2008): 527–546; Barry Rabe, "Race to the Top: The Expanding Role of U.S. State Renewable Portfolio Standards," *Sustainable Development Law and Policy* 7 (2007): 10–16; Ryan Thombs and Andrew Jorgenson, "The Political Economy of Renewable Portfolio Standards in the United States," *Energy Research and Social Science* 62 (2020): 101379.
17. National Research Council, *Changing Climate: Report of the Carbon Dioxide Assessment Committee* (Washington, DC: National Academies Press, 1983).
18. G. Nemet and E. Baker, "Demand Subsidies Versus R&D: Comparing the Uncertain Impacts of Policy on a Pre-commercial Low-Carbon Energy Technology," *Energy Journal* 30 (2009): 49–80; Lawrence Goulder and Stephen Schneider, "Induced Technological Change and the Attractiveness of CO_2 Abatement Technology," *Resource and Energy Economics* 21, no. 3–4 (1999): 211–253.
19. Sanya Carley, "State Renewable Energy Electricity Policies: An Empirical Evaluation of Effectiveness," *Energy Policy* 37 (2009): 3071–3081.
20. National Renewable Energy Laboratory (NREL), *Evaluating Renewable Portfolio Standards and Carbon Cap Scenarios in the U.S. Electric Sector*, NREL/TP-6A2-48258 (Golden, CO: NREL, 2010).

21. Prasad and Munch, "State-Level Renewable Electricity Policies"; Rabe, "Racing to the Top."
22. Paul Joskow and Richard Schmalensee, "The Performance of Coal-Burning Electric Generating Units in the United States: 1960–1980," *Journal of Applied Econometrics* 2 (1987): 85–109.
23. Lyon and Yin, "Why Do States Adopt Renewable Portfolio Standards?"; Thombs and Jorgenson, "The Political Economy of Renewable Portfolio Standards"; U.S. Energy Information Administration, *U.S. Energy-Related Carbon Emissions, 2010* (Washington, DC: U.S. Department of Energy, 2011).
24. John W. Meyer, Francisco Ramirez, and Yasemin Soysal, "World Expansion of Mass Education, 1870–1980," *Sociology of Education* 65, no. 2 (1992): 128–149; Francisco Ramirez, Yasemin Soysal, and Suzanne Shanahan, "The Changing Logic of Political Citizenship: Cross-National Acquisition of Women's Suffrage Rights, 1890 to 1990," *American Sociological Review* 62, no. 5 (1997): 735–745; Bridget Hutter and Joan O'Mahoney, *The Role of Civil Society Organisations in Regulating Business* (London: London School of Economics and Political Science, 2004).
25. David John Frank, Ann Hironaka, and Evan Schofer, "The Nation-State and the Natural Environment Over the Twentieth Century," *American Sociological Review* 65 (2000): 96–116.
26. Robert Putnam, *Bowling Alone: The Collapse and Revival of American Community* (New York: Simon & Schuster, 2000).
27. John W. Meyer and Brian Rowan, "Institutionalized Organizations: Formal Structure as Ceremony and Myth," *American Journal of Sociology* 83, no. 2 (1977): 340–363.
28. Frank, Hironaka, and Schofer, "The Nation-State and the Natural Environment"; Schofer and Hironaka, "The Effects of World Society"; John Shandra et al., "International Nongovernmental Organizations and Carbon Dioxide Emissions in the Developing World: A Quantitative, Cross-National Analysis," *Sociological Inquiry* 74 (2004): 520–544.
29. Doug McAdam et al., " 'Site Fights': Explaining Opposition to Pipeline Projects in the Developing World," *Sociological Forum* 25, no. 3 (2010): 401–427.
30. Patricia Bromley and Walter Powell, "From Smoke and Mirrors to Walking the Talk: Decoupling in the Contemporary World," *Academy of Management Annals* 6 (2012): 483–530.

31. Although closely related, energy efficiency and emission rate are conceptually distinct. When applied to power plants, the former basically refers to the fuel energy input required to generate one unit of electricity, whereas the latter refers to the pounds of CO_2 released per unit of electricity produced.
32. Karen Palmer and Anthony Paul, *A Primer on Comprehensive Policy Options for States to Comply with the Clean Power Plan* (Washington, DC: Resources for the Future, 2015).
33. Dana Fisher and Andrew Jorgenson, "Ending the Stalemate: Toward a Theory of Anthro-shift," *Sociological Theory* 37 (2019): 342–362.
34. Bromley and Powell, "From Smoke and Mirrors to Walking the Talk."
35. See also Ann Hironaka and Evan Schofer, "Loose Coupling in the Environmental Arena: The Case of Environmental Impact Assessments," in *Organizations, Policy and the Natural Environment: Institutional and Strategic Perspectives*, ed. Andrew Hoffman and Marc Ventresca (Stanford, CA: Stanford University Press, 2002).
36. John Polimeni et al., *The Jevons Paradox and the Myth of Resource Efficiency Improvements* (London: Earthscan, 2008).
37. Bruce Podobnik, *Global Energy Shifts: Fostering Sustainability in a Turbulent Age* (Philadelphia: Temple University Press, 2006), 13; cf. Ion Bogdan Vasi, *Winds of Change: The Environmental Movement and the Global Development of the Wind Energy Industry* (Oxford: Oxford University Press, 2011); Ion Bogdan Vasi, "Social Movements and Industry Development: The Environmental Movement's Impact on the Wind Energy Industry," *Mobilization* 14, no. 3 (2009): 315–336.
38. R. Cavanaugh, L. Mott, J. R. Beers, and T. Lash. Natural Resources Defense Council. *Choosing an Electrical Energy Future for the Pacific Northwest: An Alternative Scenario* (1977).
39. Paul Hirsch, *Power Loss: The Origins of Deregulation and Restructuring in the American Electric Utility System* (Cambridge, MA: MIT Press, 1999), 211.
40. Rachel Schurman and William Munro, "Targeting Capital: A Cultural Economy Approach to Understanding the Efficacy of Two Anti-genetic Engineering Movements," *American Journal of Sociology* 115, no. 1 (2009): 155–202; Ion Bogdan Vasi and Brayden King, "Social Movements, Risk Perceptions, and Economic Outcomes: The Effect of Primary and Secondary Stakeholder Activism on Firms' Perceived Environmental Risk and Financial Performance," *American Sociological Review* 77, no. 4 (2012): 573–597.

41. Edward Walsh, Rex Warland, and D. Clayton Smith, "Backyards, NIMBYs, and Incinerator Sitings: Implications for Social Movement Theory," *Social Problems* 40, no. 1 (1993): 25–38; Susan Stall and Randy Stoecker, "Community Organizing or Organizing Community?" *Gender and Society* 12, no. 6 (1998): 729–756; Benjamin Lind and Judith Stepan-Norris, "The Relationality of Movements: Movement and Countermovement Resources, Infrastructure, and Leadership in the Los Angeles Tenants' Rights Mobilization, 1976–1979," *American Journal of Sociology* 116, no. 5 (2011): 1564–1609.
42. E.g., Joseph Luders, "The Economics of Movement Success: Business Responses to Civil Rights Mobilization," *American Journal of Sociology* 111, no. 4 (2006): 963–998.
43. Robert D. Bullard, *Dumping in Dixie: Race, Class, and Environmental Quality* (Boulder, CO: Westview, 1990).
44. Silas House and Jason Howard, *Something's Rising: Appalachians Fighting Mountaintop Removal* (Lexington: University Press of Kentucky, 2009).
45. Robert D. Bullard and Glenn S. Johnson, "Environmentalism and Public Policy: Environmental Justice: Grassroots Activism and Its Impact on Public Policy Decision Making," *Journal of Social Issues* 56, no. 3 (2000): 555–578.
46. World Resources Institute (WRI), *Target: Intensity, an Analysis of Greenhouse Gas Intensity Targets* (Washington, DC: WRI, 2006).
47. Steiner Andresen and Lars Gulbrandsen, *The Role of NGOs in Promoting Climate Compliance* (Lysaker, Norway: Fridtjof Nansen Institute, 2003).
48. U.S. Environmental Protection Agency (EPA), *Available and Emerging Technologies for Reducing Greenhouse Gas Emissions from Coal-Fired Electric Generating Unit* (Washington, DC: EPA, 2010).
49. H. Selin and S. VanDeveer, "Climate Leadership in Northeast North America," in *Changing Climates in North American Politics: Institutions, Policymaking, and Multilevel Governance*, ed. H. Selin and S. VanDeveer (Cambridge, MA: MIT Press, 2009), 111–136.
50. Michaels, "National Renewable Portfolio Standard."
51. Indirect and direct climate policies are correlated .44 and .60, respectively, with the democratic state variable, suggesting that states controlled

by Democrats are more likely to adopt these measures. However, the inclusion and exclusion of the democratic state variable did not alter the effects of climate policies.

52. Population density did not significantly alter the effect of ENGOs. Neither did another possible determinant of ENGO formation, median county income, which we tested but did not report here. In addition, our two measures of industry influence exerted a nonsignificant effect when combined into a single indicator.
53. We should note that this finding is inconsistent with our analysis in chapter 4, in which total emissions were shown to increase in older plants. This is likely due to the different samples in each chapter, though more research is needed.
54. We owe a special thanks to Ion Bogdan Vasi for these analyses. Various query terms were used, such as "[power plant name] within same paragraph as [environmental protest] or [environmental demonstration] or [environmental activism] or [environmental activist] or [environmental group] or [environmentalists]."
55. Cate Lecuyer, "Lawsuit: City Plant Violated Clean Air Act Over and Over," *Salem News*, June 25, 2010.
56. Samantha Caravello, "CLF and Healthlink Win Enforceable Shutdown of Salem Harbor Station Sealing Coal Plant's Fate," Conservation Law Foundation, February 7, 2012, https://www.clf.org/newsroom/enforceable-shutdown-salem-harbor/.
57. Amanda McGregor, "Fighting the Power: Dozens Brave Cold to Protest Salem Harbor Station," *Salem News*, March 2, 2009.
58. Vasi, *Winds of Change*.
59. Vasi, *Winds of Change*.
60. Hirsch, *Power Loss*.
61. Hirsch, *Power Loss*, 220.
62. Gerald Davis and W. Richard Scott, *Organizations and Organizing: Rational, Natural, and Open Perspectives* (New York: Routledge, 2016).
63. Arthur P. J. Mol, Gert Spaargaren, and David A. Sonnenfeld, "Ecological Modernization Theory: Taking Stock, Moving Forward," in *The Routledge International Handbook of Social and Environmental Change*, ed. Stewart Lockie, David Sonnenfeld, and Dana R. Fisher (New York: Routledge, 2014), 15–30.

64. Geoff Martin and Eri Saikawa, "Effectiveness of State Climate and Energy Policies in Reducing Power-Sector CO_2 Emissions," *Nature Climate Change* 7 (2017): 912–919.
65. Lyon and Yin, "Why Do States Adopt Renewable Portfolio Standards?"
66. Neil Gunningham and Peter Grabosky, *Smart Regulation: Designing Environmental Policy* (Oxford: Clarendon Press, 1998).
67. Lawrence Goulder and Robert Stavins, "Interactions Between State and Federal Climate Change Policies" (Working Paper 16123, National Bureau of Economic Research, Cambridge, MA, 2010), www.nber.org/papers/w16123.
68. U.S. Energy Information Administration, *U.S. Energy-Related Carbon Emissions.*

6. NEXT STEPS: FUTURE RESEARCH AND ACTION ON SOCIETY'S SUPER POLLUTERS

1. "Climate Strike in Colorado," *Denver Post*, September 20, 2019.
2. Doug McAdam, "Social Movement Theory and the Prospects for Climate Change Activism in the United States," *Annual Review of Political Science* 20 (2017): 189–208.
3. See also Dana Fisher, *American Resistance: From the Women's March to the Blue Wave* (New York: Columbia University Press, 2019).
4. Mary Collins, Ian Munoz, and Joseph JaJa, "Linking 'Toxic Outliers' to Environmental Justice Communities," *Environmental Research Letters* 11 (2016): 015004.
5. Monica Prasad, "Problem-Solving Sociology," *Contemporary Sociology* 47 (2018): 393–398.
6. IPAT is an equation indicating that environmental impact (I) is a function of three factors: population (P), affluence (A), and technology (T).
7. POETIC is an extension of the IPAT formulation that attributes environmental impacts to population (P), organization (O), environment (E), technology (T), institutions (I), and culture (C).
8. David Pellow, *Garbage Wars: The Struggle for Environmental Justice in Chicago* (Cambridge, MA: MIT Press, 2002).
9. Prasad, "Problem-Solving Sociology."
10. "Canadian Coal-Fired Power Plant Transformed Into Solar Farm," *Yale Environment 360*, Yale School of Forestry and Environmental Studies,

April 8, 2019, https://e360.yale.edu/digest/canadian-nanticoke-coal-fired-power-plant-transformed-in-solar-farm.

11. Matthis Wackernagel and William Rees, *Our Ecological Footprint: Reducing Human Impact on the Earth* (Gabriola Island, BC: New Society Publishers, 1998).
12. Harland Prechel, "Corporate Power and U.S. Economic and Environmental Policy, 1978–2008," *Cambridge Journal of Regions, Economy and Society* 5 (2012): 357–375.
13. Steve Rayner, "How to Eat an Elephant: A Bottom-Up Approach to Climate Policy," *Climate Policy* 10 (2010): 615–621.
14. Charles Sabel and David G. Victor, "Governing Global Problems Under Uncertainty: Making Bottom-Up Climate Policy Work," *Climatic Change* 44, no. 1 (2017): 15–27; Charles Sabel and Jonathan Zeitlin, "Experimentalist Governance," in *The Oxford Handbook of Governance*, ed. D. Levi-Faur (New York: Oxford University Press, 2012); Jonathan Zeitlin, "EU Experimentalist Governance in Times of Crisis," *West European Politics* 39, no. 5 (2016): 1073–1094.
15. Jamie Peck and Nik Theodore, *Fast Policy: Experimental Statecraft at the Thresholds of Neoliberalism* (Minneapolis: University of Minnesota Press, 2015).
16. Harland Prechel and Alesha Istvan, "Disproportionality of Corporations' Environmental Pollution in the Electrical Energy Industry," *Sociological Perspectives* 59, no. 3 (2016): 505–527.
17. Julie Battilana and Silvia Dorado, "Building Sustainable Hybrid Organizations: The Case of Commercial Micro-Finance Organizations," *Academy of Management Journal* 6 (2010): 1419–1440.
18. Steven Lewis and Anna Hogan, "Reform First and Ask Questions Later?: The Implications of (Fast) Schooling Policy and 'Silver Bullet' Solutions," *Critical Studies in Education* 60, no. 1 (2019): 1–18.
19. Richard Heede, "Tracing Anthropocentric Carbon Dioxide and Methane Emissions to Fossil Fuel and Cement Producers, 1854–2010," *Climatic Change* 122, no. 1–2 (2014): 229–241; CDP, *CDP Carbon Majors Report 2017* (London: CDP, 2017), https://6fefcbb86e61af1b2fc4-c70d8ead6ced550b4d987d7c03-fcdd1d.ssl.cf3.rackcdn.com/cms/reports/documents/000/002/327/original/Carbon-Majors-Report-2017.pdf?1501833772.
20. Dario Kenner, *Carbon Inequality: The Role of the Richest in Climate Change* (Abingdon, UK: Routledge, 2019); Oxfam, *Extreme Carbon Inequality*,

Media Briefing (Nairobi: Oxfam, December 2, 2015), https://www-cdn.oxfam.org/s3fs-public/file_attachments/mb-extreme-carbon-inequality-021215-en.pdf.

21. Lucas Chancel and Thomas Piketty, *Carbon and Inequality: From Kyoto to Paris* (Pars: Paris School of Economics, 2015).
22. B. Ekwurzel et al., "The Rise in Global Atmospheric CO_2, Surface Temperature, and Sea Level from Emissions Traced to Major Carbon Producers," *Climatic Change* 144 (2017): 579–590.
23. Chancel and Piketty, "Carbon and Inequality."
24. Emily Atkin, "The Growing Movement to Take Polluters to Court Over Climate Change," *New Republic*, December 20, 2017.
25. Energy and Policy Institute, *Utilities Knew: Documenting Electric Utilities' Early Knowledge and Ongoing Deception of Climate Change from 1969–2017* (San Francisco: Energy Policy Institute, 2017).
26. "Poll: Majority in All States, Congressional Districts Support Clean Power Plan," Yale School of Forestry and Environmental Studies, 2016, https://environment.yale.edu/news/article/poll-majority-support-for-clean-power-plan-in-all-states-congressional-districts/.
27. An obvious exception is fracking sites, some of which leak hugely disproportionate amounts of invisible methane gas. Jonah Kessel and Hiroko Tabuchi, "It's a Vast, Invisible Climate Menace. We Made It Visible," *New York Times*, December 12, 2019.
28. Marjan Minnesma, "Not Slashing Emissions? See You in Court," *Nature* 576 (2019): 379–382.
29. A major issue in these types of decisions is whether the burning of fossil fuels constitutes a public nuisance and, if so, whether judges have to compare the benefits of burning fossil fuels to the harms from climate change.
30. Matt Kempner, "Never Say Never: Atlanta Draws Fresh Jolt of Energy from Sunshine," *Atlanta Journal-Constitution*, February 15, 2019.
31. McAdam, "Social Movement Theory."
32. Ion Bogdan Vasi and Brayden King, "Social Movements, Risk Perceptions, and Economic Outcomes: The Effect of Primary and Secondary Stakeholder Activism on Firms' Perceived Environmental Risk and Financial Performance," *American Sociological Review* 77, no. 4 (2012): 573–597.
33. Carsten Q. Schneider and Claudius Wagemann, "Reducing Complexity in Qualitative Comparative Analysis (QCA): Remote and Proximate

Factors and the Consolidation of Democracy," *European Journal of Political Research* 45, no. 5 (2006): 751–786.

34. "Colorado May Have a Winning Formula for Managing Early Coal Plant Retirements," Green Tech Media, March 28, 2019, https://www.greentechmedia.com/articles/read/colorado-may-have-a-winning-formula-for-managing-early-coal-plant-retiremen#gs.sccav1.
35. Paul Krugman, "The Big Green Test: Conservatives and Climate Change," *New York Times*, June 23, 2014.
36. Most experts agree that the true social cost of carbon is approximately $50 per ton, although others estimate that it should be considerably higher or near $250 per ton. See Noah Kaufman and Kate Gordon, *The Energy, Economic, and Emissions Impacts of a Federal US Carbon Tax* (New York: Columbia Center on Global Energy Policy, 2018); Chelsea Harvey, "Should the Social Cost of Carbon Be Higher?" *E&E News*, November 22, 2017.
37. Glen Peters et al., "Growth in Emission Transfers Via International Trade from 1990 to 2008," *Proceedings of the National Academy of Sciences* 108, no. 21 (2011): 8903–8908.
38. Marc Hafstead et al., "Macroeconomic Analysis of Federal Carbon Taxes" (Policy Brief 16-06, Resources for the Future, Washington, DC, June 2016).
39. To minimize the financial harm such a tax might do to low-income customers and employees of super polluter plants, an income tax rebate and/or payroll tax rebate would likely be the most effective. Terry Dinan, "Offsetting a Carbon Tax's Costs on Low-Income Households" (Working Paper 2012-16, Congressional Budget Office, Washington, DC, 2012).
40. Intergovernmental Panel on Climate Change (IPCC), *Global Warming of 1.5°C* (Geneva: IPCC, 2018), https://www.ipcc.ch/sr15/download/.
41. "Mapped: The World's Coal Power Plants," Carbon Brief, March 26, 2020, https://www.carbonbrief.org/mapped-worlds-coal-power-plants.
42. International Energy Agency (IEA), *Coal 2017: Analysis and Forecasts to 2022* (Paris: IEA, 2017), https://www.iea.org/coal2017/.
43. Dan Tong, Qiang Zhang, Steve J. Davis, Fei Liu, Bo Zheng, Guannan Geng, Tao Xue, et al., "Targeted Emission Reductions from Global Super-Polluting Power Plant Units," *Nature Sustainability* 1 (2018): 59.
44. Collins, Munoz, and JaJa, "Linking 'Toxic Outliers.' "

45. Charles Perrow, "Organizations and Global Warming," in *Routledge Handbook of Climate Change and Society*, ed. Constance Lever-Tracy (New York: Routledge, 2010), 59–77.

APPENDIX TO CHAPTER 2

1. Eugene Rosa and Thomas Dietz, "Human Drivers of National Greenhouse Gas Emissions," *Nature Climate Change* 2 (2012): 581–586.
2. Riley Dunlap and Robert J. Brulle, eds., *Climate Change and Society: Sociological Perspectives* (New York: Oxford University Press, 2015); Timmons Roberts and Bradley Parks, *A Climate of Injustice: Global Inequality, North-South Politics, and Climate Policy* (Cambridge, MA: MIT Press, 2007).
3. Andrew Jorgenson et al., "Social Science Perspectives on Drivers of and Responses to Global Climate Change," *Wiley Interdisciplinary Reviews: Climate Change* 10, no. 1 (2019): e554, https://doi.org/10.1002/wcc.554.
4. Thomas Dietz, Scott Frey, and Linda Kalof, "Estimation with Cross-National Data: Robust and Resampling Estimators," *American Sociological Review* 52 (1987): 380–390.
5. Larry Hamilton, *Regression with Graphics* (Pacific Grove, CA: Duxbury, 1992).

APPENDIX TO CHAPTER 3

1. Benoit Rihoux and Charles Ragin, eds., *Configurational Comparative Methods: Qualitative Comparative Analysis (QCA) and Related Techniques* (Thousand Oaks, CA: Sage, 2009).
2. Peer C. Fiss, "Building Better Causal Theories: A Fuzzy Set Approach to Typologies in Organization Research," *Academy of Management Journal* 54 (2011): 393–420.
3. Rob Clark and Jason Beckfield, "A New Trichotomous Measure of World-System Position Using the International Trade Network," *International Journal of Comparative Sociology* 50 (2009): 5–38.
4. Wesley Longhofer et al., "NGOs, INGOs, and Social Change: Environmental Policy Reform in the Developing World, 1970–2010," *Social Forces* 94 (2016): 1743–1768.

5. Mario E. Bergara, Witold J. Henisz, and Pablo T. Spiller, "Political Institutions and Electric Utility Investment: A Cross-Nation Analysis," *California Management Review* 40 (1998): 18–35.
6. Charles Perrow, "Organizations and Global Warming," in *Routledge Handbook of Climate Change and Society*, ed. Constance Lever-Tracy (New York: Routledge), 59–77.
7. Andrew Jorgenson and Brett Clark, "Are the Economy and the Environment Decoupling?: A Comparative International Study, 1960–2005," *American Journal of Sociology* 118 (2012): 1–44; Richard York, Eugene Rosa, and Thomas Dietz, "Footprints on the Earth: The Environmental Consequences of Modernity," *American Sociological Review* 68 (2003): 279–300.
8. Eugene Rosa and Thomas Dietz, "Human Drivers of National Greenhouse Gas Emissions," *Nature Climate Change* 2 (2012): 581–586.
9. Charles C. Ragin, *Fuzzy-Set Social Science* (Chicago: University of Chicago Press, 2000).
10. Ragin, *Fuzzy-Set Social Science.*
11. Ragin, *Fuzzy-Set Social Science.*
12. U.S. Environmental Protection Agency (EPA), *Available and Emerging Technologies for Reducing Greenhouse Gas Emissions from Coal-Fired Electric Generating Units* (Washington, DC: EPA, 2010).
13. Ragin, *Fuzzy-Set Social Science.*
14. Kyle Longest and Stephen Vaisey, "Fuzzy: A Program for Performing Qualitative Comparative Analyses (QCA) in Stata," *Stata Journal* 8 (2008): 79–104.
15. Charles C. Ragin and Peer C. Fiss, "Net Effects Versus Configurations: An Empirical Demonstration," in *Redesigning Social Inquiry: Fuzzy Sets and Beyond*, ed. Charles C. Ragin (Chicago: University of Chicago Press, 2008), 190–212; Thomas Greckhamer, Vilmos Misangyi, and Peer Fiss, "The Two QCAs: From a Small-N to a Large-N Set Theoretic Approach," *Research in Sociology of Organizations* 58 (2013): 51–77.
16. Sophia Rabe-Hesketh and Anders Skrondal, *Multilevel and Longitudinal Modeling Using Stata*, 2nd ed. (College Station, TX: Stata Press, 2008).
17. Greckhamer, Misangyi, and Fiss, "The Two QCAs."
18. Vincent Roscigno and Randy Hodson, "The Organizational and Social Foundations of Worker Resistance," *American Sociological Review* 69

(2004): 14–39; Don Grant et al., "Bringing the Polluters Back In: Environmental Inequality and the Organization of Chemical Production," *American Sociological Review* 75 (2010): 479–504.

19. Jerry M. Mendel and Charles C. Ragin, "fsQCA: Dialog Between Jerry M. Mendel and Charles C. Ragin" (USC-SIPI Report 411, University of Southern California, Los Angeles, 2011), 9, http://sipi.usc.edu/reports/pdfs/Originals/USC-SIPI-411.pdf.
20. Peer C. Fiss, Dmitry Sharapov, and Lasse Cronqvist, "Opposites Attract?: Opportunities and Challenges for Integrating Large-N QCA and Econometric Analysis," *Political Research Quarterly* 66 (2013): 191–198.
21. Mendel and Ragin, "fsQCA."
22. Greckhamer, Misangyi, and Fiss, "The Two QCAs."

APPENDIX TO CHAPTER 4

1. David Wheeler and Kevin Ummel, "Calculating CARMA: Global Estimation of CO_2 Emissions from the Power Sector," Working Paper 145, Center for Global Development, Washington, DC, https://www.cgdev.org/publication/calculating-carma-global-estimation-co2-emissions-power-sector-working-paper-145.
2. Rob Clark and Jason Beckfield, "A New Trichotomous Measure of World-System Position Using the International Trade Network," *International Journal of Comparative Sociology* 50 (2009): 5–38.
3. Wesley Longhofer and Evan Schofer, "National and Global Origins of Environmental Association," *American Sociological Review* 75 (2010): 505–533.
4. Eugene Rosa and Thomas Dietz, "Human Drivers of National Greenhouse Gas Emissions," *Nature Climate Change* 2 (2012): 581–586.
5. Andrew Jorgenson and Brett Clark, "Societies Consuming Nature: A Panel Study of the Ecological Footprints of Nations, 1960–2003," *Social Science Research* 40 (2011): 226–244; Richard York, Eugene Rosa, and Thomas Dietz, "Footprints on the Earth: The Environmental Consequences of Modernity," *American Sociological Review* 68 (2003): 279–300.

APPENDIX TO CHAPTER 5

1. Wesley Sine and Brandon Lee, "Tilting at Windmills?: The Environmental Movement and the Emergence of the U.S. Wind Energy Sector," *Administrative Science Quarterly* 54 (2009): 123–155.
2. Don Grant, Andrew Jones, and Mary Nell Trautner, "Do Facilities with Distant Headquarters Pollute More?: How Civic Engagement Conditions the Environmental Performance of Absentee Managed Plants," *Social Forces* 83, no. 1 (2004): 189–214; Charles Tolbert, Thomas Lyson, and Michael Irwin, "Local Capitalism, Civic Engagement, and Socioeconomic Well-Being," *Social Forces* 77 (1998): 401–428.

REFERENCES

Adair, Sarah, David Hoppock, and Jonas Monast. 2014. "New Source Review and Coal Plant Efficiency Gains: How New and Forthcoming Air Regulations Affect Outcomes." *Energy Policy* 70: 183–192.

Andresen, Steiner, and Lars Gulbrandsen. 2003. *The Role of NGOs in Promoting Climate Compliance*. Lysaker, Norway: Fridtjof Nansen Institute.

Antonio, Robert, and Brett Clark. 2015. "The Climate Change Divide in Social Theory." In *Climate Change and Society: Sociological Perspectives*, ed. Riley Dunlap and Robert Brulle, 333–368. New York: Oxford University Press.

Ard, Kerry. 2015. "Trends in Exposure to Industrial Air Toxins for Different Racial and Socioeconomic Groups: A Spatial and Temporal Examination of Environmental Inequality in the U.S. from 1995 to 2004." *Social Science Research* 53: 375–390.

Armstrong, Elizabeth, and Laura Hamilton. 2013. *Paying for the Party: How College Maintains Inequality*. Cambridge, MA: Harvard University Press.

Asia-Pacific Economic Cooperation (APEC), Energy Working Group. 2005. *Costs and Effectiveness of Upgrading and Refurbishing Older Coal-Fired Power Plants in Developing APEC Economies*. Singapore: APEC. www.egcfe.ewg.apec.org/projects/UpgradePP_Report_2005.pdf.

Atkin, Emily. 2017. "The Growing Movement to Take Polluters to Court Over Climate Change." *New Republic*. December 20.

Battilana, Julie, and Silvia Dorado. 2010. "Building Sustainable Hybrid Organizations: The Case of Commercial Micro-Finance Organizations." *Academy of Management Journal* 6: 1419–1440.

Beér, Janos. 2006. "High Efficiency Electric Power Generation: The Environmental Role." *Progress in Energy Combustion Science* 33, no. 2: 107–134.

Bergara, Mario E., Witold J. Henisz, and Pablo T. Spiller. 1998. "Political Institutions and Electric Utility Investment: A Cross-Nation Analysis." *California Management Review* 40: 18–35.

Berger, Peter L., and Thomas Luckman. 1967. *The Social Construction of Reality*. New York: Doubleday.

Bernstein, Steven, and Matthew Hoffman. "Climate Politics, Metaphors and the Fractal Carbon Trap." *Nature Climate Change* 9 (2019): 919–925.

Brecher, Jeremy. 2015. *Climate Insurgency: A Strategy for Survival*. New York: Routledge.

Bromley, Patricia, and Walter Powell. 2012. "From Smoke and Mirrors to Walking the Talk: Decoupling in the Contemporary World." *Academy of Management Annals* 6: 483–530.

Brookes, Len. 1979. "A Low-Energy Strategy for the UK by G. Leach et al.: A Review and Reply." *Atom* 269: 3–8.

Brulle, Robert, and David Pellow. 2006. "Environmental Justice: Human Health and Environmental Inequalities." *Annual Review of Public Health* 27: 103–124.

Bullard, Robert. 1983. "Solid Waste Sites and the Black Houston Community." *Sociological Inquiry* 53: 273–288.

Bullard, Robert. 1990. *Dumping in Dixie: Race, Class, and Environmental Quality*. Boulder, CO: Westview.

Bullard, Robert D., and Glenn S. Johnson. 2000. "Environmentalism and Public Policy: Environmental Justice: Grassroots Activism and Its Impact on Public Policy Decision Making." *Journal of Social Issues* 56, no. 3: 555–578.

Bunker, Stephen, and Paul Ciccantell. 2005. *Globalization and the Race for Resources*. Baltimore: Johns Hopkins University Press.

Burawoy, Michael. 2005. "For Public Sociology." *American Sociological Review* 70 no. 1: 4–28.

Burns, Thomas, Byron Davis, and Edward Kick. 1997. "Position in the World-System and National Emissions of Greenhouse Gases." *Journal of World-Systems Research* 3, no. 3: 432–466.

Bushnell, James, Carla Peterman, and Catherine Wolfram. 2008. "Local Solutions to Global Problems: Climate Change Policies and Regulatory Jurisdiction." *Review of Environmental Economics and Policy* 2, no. 2: 175–193.

Buttel, Frederick. 1987. "New Directions in Environmental Sociology." *Annual Review of Sociology* 13: 465–488.

Buttel, Frederick H. 2000. "World Society, the Nation-State, and Environmental Protection: Comment on Frank, Hironaka, and Schofer." *American Sociological Review* 65: 117–121.

Buttel, Frederick H. 2004. "The Treadmill of Production: An Appreciation, Assessment, and Agenda for Research." *Organization and Environment* 17: 323–336.

Campbell, Richard. 2013. *Increasing the Efficiency of Existing Coal-Fired Power Plants*. Washington, DC: Congressional Research Service. https://fas.org/sgp/crs/misc/R43343.pdf.

Caravello, Samantha, "CLF and Healthlink Win Enforceable Shutdown of Salem Harbor Station Sealing Coal Plant's Fate," Conservation Law Foundation, February 7, 2012, https://www.clf.org/newsroom/enforceable-shutdown-salem-harbor/.

Carley, Sanya. 2009. "State Renewable Energy Electricity Policies: An Empirical Evaluation of Effectiveness." *Energy Policy* 37: 3071–3081.

Carley, Sanya. 2011. "The Era of State Energy Policy Innovation: A Review of Policy Instruments." *Review of Policy Research* 28: 265–294.

Castesana, Paula, and Salvador Puliafito. 2014. "Development of an Agent-Based Model and Its Application to the Estimation of Global Carbon Emissions." *Low Carbon Economy* 4: 24–34.

Catton, William, and Riley Dunlap. 1978. "Environmental Sociology: A New Paradigm." *American Sociologist* 13: 41–49.

Cavanaugh, R., L. Mott, J. R. Beers, and T. Lash. Natural Resources Defense Council. *Choosing an Electrical Energy Future for the Pacific Northwest: An Alternative Scenario* (1977).

Cavlovic, Therese A., Kenneth H. Baker, Robert P. Berrens, and Kishore Gawande. 2000. "A Meta-analysis of Environmental Kuznets Curve Studies." *Journal of Industrial Ecology* 4, no. 4: 13–29.

CDP. 2017. *CDP Carbon Majors Report 2017*. London: CDP. https://6fefcbb86-e61af1b2fc4-c70d8ead6ced550b4d987d7c03fcdd1d.ssl.cf3.rackcdn.com/cms/reports/documents/000/002/327/original/Carbon-Majors-Report-2017.pdf?1501833772.

Center for Clean Air Policy (CCAP). 2008. *Sectoral Approaches: A Pathway to Nationally Appropriate Mitigation Actions*. Washington, DC: CCAP. http://ccap.org/assets/Center-for-Clean-Air-Policy-Interim-Report-Sectoral

-Approaches-A-Pathway-to-Nationally-Appropriate-Mitigation-Actions_CCAP-December-20081.pdf.

Chancel, Lucas, and Thomas Piketty. 2015. *Carbon and Inequality: From Kyoto to Paris*. Paris: Paris School of Economics.

Chase-Dunn, Christopher. 1998. *Global Formation: Structures of the World Economy*. Lanham, MD: Rowman & Littlefield.

Chase-Dunn, Christopher, and Peter Grimes. 1995. "World-Systems Analysis." *Annual Review of Sociology* 21: 387–417.

Choudry, Azis, and Dip Kapoor, eds. 2013. *NGOization: Complicity, Contradictions and Prospects*. London: Zed Books.

Churchman, C. West. 1967. "Guest Editorial: Wicked Problems." *Management Science* 14, no. 4: B141–B142.

Clark, Brett, and Richard York. 2005. "Carbon Metabolism: Global Capitalism, Climate Change, and the Biospheric Rift." *Theory and Society* 34, no. 4: 391–428.

Clark, Rob, and Jason Beckfield. 2009. "A New Trichotomous Measure of World-System Position Using the International Trade Network." *International Journal of Comparative Sociology* 50: 5–38.

Climatewire. 2019. "Behemoth Coal Plants Threaten Utilities' CO2 Goals." August 23. https://www.eenews.net/stories/1061039369.

Cole, Wade. 2015. "Mind the Gap: State Capacity and the Implementation of Human Rights Treaties." *International Organization* 69: 405–441.

Collins, Mary. 2011. "Risk-Based Targeting: Identifying Disproportionalities in the Sources and Effects of Industrial Pollution." *American Journal of Public Health* 101: 231–237.

Collins, Mary, Ian Munoz, and Joseph JaJa. 2016. "Linking 'Toxic Outliers' to Environmental Justice Communities." *Environmental Research Letters* 11: 015004.

Crilly, Donal. 2013. "Corporate Social Responsibility: A Multilevel Explanation of Why Managers Do Good." In *Configurational Theory and Methods in Organizational Research*, ed. P. C. Fiss, B. Cambre, and A. Marx, 181–204. Bingley, UK: Emerald Group Publishing Limited.

Denver Post. 2019. "Climate Strike in Colorado." September 20.

Desmond, Matthew. 2016. *Evicted: Poverty and Profit in the American City*. New York: Broadway Books.

Diamond, Jared. 2005. *Collapse: How Societies Choose to Fail or Succeed*. New York: Viking Press.

Dietz, Thomas. 2013. "Context Matters: Eugene A. Rosa's Lessons for Structural Human Ecology." In *Structural Human Ecology: New Essays in Risk, Energy, and Sustainability*, ed. Thomas Dietz and Andrew Jorgenson, 189–215. Pullman: Washington State University Press.

Dietz, Thomas. 2015. "Prolegomenon to a Structural Human Ecology of Human Well-Being." *Sociology of Development* 1: 123–148.

Dietz, Thomas. 2017. "Drivers of Human Stress on the Environment in the Twenty-First Century." *Annual Review of Environment and Resources* 42: 189–213.

Dietz, Thomas, Kenneth A. Frank, Cameron T. Whitley, Jennifer Kelly, and Rachel Kelly. 2015. "Political Influences on Greenhouse Gas Emissions from US States." *Proceedings of the National Academy of Sciences* 112: 8254–8259.

Dietz, Thomas, Scott Frey, and Linda Kalof. 1987. "Estimation with Cross-National Data: Robust and Resampling Estimators." *American Sociological Review* 52: 380–390.

Dietz, Thomas, Gerald T. Gardner, Jonathan Gilligan, Paul C. Stern, and Michael P. Vandenbergh. 2009. "Household Actions Can Provide a Behavioral Wedge to Rapidly Reduce U.S. Carbon Emissions." *Proceedings of the National Academy of Sciences* 106: 18452–18456.

Dietz, Thomas, Eugene A. Rosa, and Richard York. 2010. "Human Driving Forces of Global Change: Dominant Perspectives." In *Human Footprints on the Global Environment: Threats to Sustainability*, ed. E. Rosa, A. Diekmann, T. Dietz, and C. Jaeger, 83–134. Cambridge, MA: MIT Press.

Dietz, Thomas, Rachael L. Shwom, and Cameron T. Whitley. Forthcoming. "Climate Change and Society." *Annual Review of Sociology*.

DiMaggio, Paul J., and Walter W. Powell. 1983. "The Iron Cage Revisited: Institutional Isomorphism and Collective Rationality in Organizational Fields." *American Sociological Review* 48: 147–160.

Dinan, Terry. 2012. "Offsetting a Carbon Tax's Costs on Low-Income Households." Working Paper 2012-16, Congressional Budget Office, Washington, DC.

Doremus, Holly, and W. Michael Hanemann. 2008. "Babies and Bathwater: Why the Clean Air Act's Cooperative Federalism Framework Is Useful for Addressing Global Warming." *Arizona Law Review* 50: 799–834.

Dunlap, Riley, and Robert J. Brulle, eds. 2015. *Climate Change and Society: Sociological Perspectives*. New York: Oxford University Press.

Egenhofer, Christian, and Noriko Fujiwara. 2008. *Global Sectoral Industry Approaches to Global Climate Change: The Way Forward.* Brussels: Centre for European Policy Studies Task Force. https://www.ceps.eu/ceps-publications/global-sectoral-industry-approaches-climate-change-way-forward/.

Eilpern, Juliet, Brady Dennis, and Chris Mooney. 2018. "Trump Administration Sees a 7-Degree Rise in Global Temperatures by 2100." *Washington Post*. September 28.

Ekwurzel, B., J. Boneham, M. W. Dalton, R. Heede, R. J. Mera, M. R. Allen, and P. C. Frumhoff. 2017. "The Rise in Global Atmospheric CO_2, Surface Temperature, and Sea Level from Emissions Traced to Major Carbon Producers." *Climatic Change* 144: 579–590.

Energy and Policy Institute (EPI). 2017. *Utilities Knew: Documenting Electric Utilities' Early Knowledge and Ongoing Deception of Climate Change from 1969–2017*. San Francisco: EPI.

Engels, Frederick. 1892. *The Condition of the Working Class in England in 1844*. London: Sonnenschein.

Evans, Peter. 1995. *Embedded Autonomy: States and Industrial Transformation*. Princeton, NJ: Princeton University Press.

Farrell, Justin. 2016. "Network Structure and Influence of the Climate Change Counter-Movement." *Nature Climate Change* 6, no. 4: 370–374.

Fisher, Dana. 2010. "COP-15 in Copenhagen: How the Merging of Movements Left Civil Society Out in the Cold." *Global Environmental Politics* 10, no. 2: 11–17.

Fisher, Dana. 2019. *American Resistance: From the Women's March to the Blue Wave*. New York: Columbia University Press.

Fisher, Dana R., and William R. Freudenburg. 2004. "Post Industrialization and Environmental Quality: An Empirical Analysis of the Environmental State." *Social Forces* 83: 157–188.

Fisher, Dana, and Andrew Jorgenson. 2019. "Ending the Stalemate: Toward a Theory of Anthro-shift." *Sociological Theory* 37: 342–362.

Fisher, Dana, Anya Galli Robertson, Joseph McCartney Waggle, Amanda Dewey, Ann Dubin, and William Yagatich. 2018. "Polarizing Climate Politics in America." *Research in Political Sociology* 25: 1–23.

Fiss, Peer C. 2007. "A Set-Theoretic Approach to Organizational Configurations." *Academy of Management Review* 32: 1180–1198.

Fiss, Peer C. 2011. "Building Better Causal Theories: A Fuzzy Set Approach to Typologies in Organization Research." *Academy of Management Journal* 54: 393–420.

Fiss, Peer C., Bart Cambré, and Axel Marx, eds. 2013. *Configurational Theory and Methods in Organizational Research.* Bingley, UK: Emerald Group.

Fiss, Peer C., Dmitry Sharapov, and Lasse Cronqvist. 2013. "Opposites Attract?: Opportunities and Challenges for Integrating Large-N QCA and Econometric Analysis." *Political Research Quarterly* 66: 191–198.

Fligstein, Neil, and Doug McAdam. 2015. *A Theory of Fields.* New York: Oxford University Press.

Foster, John Bellamy. 1999. "Marx's Theory of Metabolic Rift." *American Journal of Sociology* 105, no. 2: 366–405.

Foster, John Bellamy, Brett Clark, and Richard York. 2010. *The Ecological Rift: Capitalism's War on the Earth.* New York: Monthly Review Press.

Foster, John Bellamy, and Hannah Holleman. 2012. "Weber and the Environment: Classical Foundations for a Postmaterialist Sociology." *American Journal of Sociology* 117, no. 6: 1625–1673.

Frank, David John, Ann Hironaka, and Evan Schofer. 2000. "The Nation-State and the Natural Environment Over the Twentieth Century." *American Sociological Review* 65: 96–116.

Freudenburg, William. 2005. "Privileged Access, Privileged Accounts: Toward a Socially Structured Theory of Resources and Discourses." *Social Forces* 84: 89–114.

Freudenburg, William. 2006. "Environmental Degradation, Disproportionality, and the Double Diversion: Reaching Out, Reaching Ahead, and Reaching Beyond." *Rural Sociology* 71, no. 1: 3–32.

Garrett, T. J. 2012. "No Way Out?: The Double-Bind in Seeking Global Prosperity Alongside Mitigated Climate Change." *Earth System Dynamics* 3: 1–17.

Gillingham, Kenneth, David Rapson, and Gernot Wagner. 2016. "The Rebound Effect and Energy Efficiency Policy." *Review of Environmental Economics and Policy* 10, no. 1: 68–88.

Giugni, Marco. 1998. "Was It Worth the Effort?: The Outcomes and Consequences of Social Movements." *Annual Review of Sociology* 24: 371–393.

Givens, Jennifer. 2017. "World Society, World Polity, and the Carbon Intensity of Well-Being, 1990–2011." *Sociology of Development* 3: 403–435.

Givens, Jennifer, and Andrew Jorgenson. 2013. "Individual Environmental Concern in the World Polity." *Social Science Research* 42: 418–431.

Goffman, Alice. 2015. *On the Run: Fugitive Life in an American City*. New York: Picador.

Gould, Kenneth, David Pellow, and Allan Schnaiberg. 2008. *The Treadmill of Production: Injustice and Unsustainability in the Global Economy*. Boulder, CO: Paradigm.

Goulder, Lawrence, and Stephen Schneider. 1999. "Induced Technological Change and the Attractiveness of CO_2 Abatement Technology." *Resource and Energy Economics* 21: 211–253.

Goulder, Lawrence, and Robert Stavins. 2010. "Interactions Between State and Federal Climate Change Policies." Working Paper 16123, National Bureau of Economic Research, Cambridge, MA. www.nber.org/papers/w16123.

Graedel, T. E., and Braden R. Allensby. 2003. *Industrial Ecology*. 2nd ed. Englewood Cliffs, NJ: Prentice Hall.

Granovetter, Mark. 1985. "Economic Structure and Social Action: The Problem of Embeddedness." *American Journal of Sociology* 91, no. 3: 481–510.

Grant, Don, Albert Bergesen, and Andrew Jones. 2002. "Organizational Size and Pollution: The Case of the U.S. Chemical Industry." *American Sociological Review* 67: 389–407.

Grant, Don, Andrew Jones, and Mary Nell Trautner. 2004. "Do Facilities with Distant Headquarters Pollute More?: How Civic Engagement Conditions the Environmental Performance of Absentee Managed Plants." *Social Forces* 83, no. 1: 189–214.

Grant, Don, Mary Nell Trautner, Liam Downey, and Lisa Thiebaud. 2010. "Bringing the Polluters Back In: Environmental Inequality and the Organization of Chemical Production." *American Sociological Review* 75: 479–504.

Greckhamer, Thomas, Vilmos Misangyi, and Peer Fiss. 2013. "The Two QCAs: From a Small-N to a Large-N Set Theoretic Approach." *Research in Sociology of Organizations* 58: 51–77.

Greening, Lorna, David Greene, and Carmen Difiglio. 2000. "Energy Efficiency and Consumption—The Rebound Effect—A Survey." *Energy Policy* 28, no. 6: 389–401.

Greenwald, Bruce, and Joseph Stiglitz. 1986. "Externalities in Economies with Imperfect Information and Incomplete Markets." *Quarterly Journal of Economics* 101: 229–264.

Greenwald, Bruce, and Joseph Stiglitz. 1993. "New and Old Keynesians." *Journal of Economic Perspectives* 7, no. 1: 23–44.

Gunderson, Ryan. "Explaining Technological Impacts Without Determinism: Fred Cottrell's Sociology of Technology and Energy." *Energy Research and Social Science* 42 (2018): 127–133.

Gunningham, Neil, and Peter Grabosky. 1998. *Smart Regulation: Designing Environmental Policy*. Oxford: Clarendon Press.

Hafstead, Marc, Lawrence Goulder, Raymond Kopp, and Roberton Williams III. 2016. "Macroeconomic Analysis of Federal Carbon Taxes." Policy Brief 16-06, Resources for the Future, Washington, DC, June.

Hamilton, Larry. 1992. *Regression with Graphics*. Pacific Grove, CA: Duxbury.

Hannan, Michael T., and John Freeman.1984. "Structural Inertia and Organizational Change." *American Sociological Review* 49, no. 2: 149–164.

Harvey, Chelsea. 2017. "Should the Social Cost of Carbon Be Higher?" *E&E News*. November 22.

Heede, Richard. 2014. "Tracing Anthropocentric Carbon Dioxide and Methane Emissions to Fossil Fuel and Cement Producers, 1854–2010." *Climatic Change* 122, no. 1–2: 229–241.

Henisz, Witold J. 2002. "The Institutional Environment for Infrastructure Investment." *Industrial and Corporate Change* 11: 355–389.

Henisz, Witold J., Bennet A. Zelner, and Mauro F. Guillén. 2005. "The Worldwide Diffusion of Market-Oriented Infrastructure Reform, 1977–1999." *American Sociological Review* 70: 871–897.

Herendeen, Robert, and Jerry Tanaka. 1976. "Energy Cost of Living." *Energy* 1: 165–178.

Hilgartner, Stephen, and Charles L. Bosk. 1988. "The Rise and Fall of Social Problems: A Public Arenas Model." *American Journal of Sociology* 94, no. 1: 53–78.

Hironaka, Ann. 2014. *Greening the Globe: World Society and Environmental Change*. New York: Cambridge University Press.

Hironaka, Ann, and Evan Schofer. 2002. "Loose Coupling in the Environmental Arena: The Case of Environmental Impact Assessments." In *Organizations, Policy and the Natural Environment: Institutional and Strategic Perspectives*, ed. Andrew Hoffman and Marc Ventresca. Stanford, CA: Stanford University Press.

Hirsch, Paul. 1999. *Power Loss: The Origins of Deregulation and Restructuring in the American Electric Utility System*. Cambridge, MA: MIT Press.

Hirsch, Paul, Stuart Michaels, and Ray Friedman. 1987. " 'Dirty Hands' Versus 'Clean Models': Is Sociology in Danger of Being Seduced by Economics?" *Theory and Society* 16, no. 3: 317–336.

House, Silas, and Jason Howard. 2009. *Something's Rising: Appalachians Fighting Mountaintop Removal.* Lexington: University Press of Kentucky.

Hutter, Bridget, and Joan O'Mahoney. 2004. *The Role of Civil Society Organisations in Regulating Business.* London: London School of Economics and Political Science.

Intergovernmental Panel on Climate Change (IPCC). 2014. *Climate Change 2014: Synthesis Report.* Geneva: IPCC. https://www.ipcc.ch/report/ar5/wg2/; trillionthtonne.org.

Intergovernmental Panel on Climate Change (IPCC). 2018. *Global Warming of 1.5°C.* Geneva: IPCC. https://www.ipcc.ch/sr15/download/.

International Energy Agency (IEA). 2009. *How the Energy Sector Can Deliver on a Climate Agreement in Copenhagen.* Paris: IEA.

International Energy Agency (IEA). 2009. *Sectoral Approaches in Electricity: Building Bridges to a Safe Climate.* Paris: IEA.

International Energy Agency (IEA). 2009. *World Energy Outlook.* Paris: IEA.

International Energy Agency (IEA). 2010. *Upgrading and Efficiency Improvements in Coal-Fired Power Plants.* Paris: IEA. http://iea-coal.org/uk/site/2010/publications-section/reports.

International Energy Agency (IEA). 2012. *Energy Technology Perspectives.* Paris: OECD/IEA.

International Energy Agency (IEA). 2017. *Coal 2017: Analysis and Forecasts to 2022.* Paris: IEA. https://www.iea.org/coal2017/.

International Energy Agency (IEA). 2019. *Global Energy and CO_2 Status Report 2018.* Paris: IEA.

Jackson, Robert, C. Le Quéré, R. M. Andrew, J. G. Canadell, J. I. Korsbakken, Z. Liu, G. P. Peters, and B. Zheng. "Global Energy Growth Is Outpacing Decarbonization." *Environmental Research Letters* 13, no. 12. https://doi.org/10.1088/1748-9326/aaf303.

Javeline, Debra. 2014. "The Most Important Topic Political Scientists Are Not Studying: Adapting to Climate Change." *Perspectives on Politics* 12, no. 2: 420–434.

Jenner, Steffen, Gabriel Chan, Rolf Frankenberger, and Mathias Gabel. 2012. "What Drives States to Support Renewable Energy?" *Energy Journal* 33: 1–12.

Jevons, William Stanley. 1965. *The Coal Question: An Inquiry Concerning the Progress of the Nation, and the Probable Exhaustion of Our Coal-Mines.* 3rd ed., ed. A. W. Flux. New York: Augustus M. Kelley. First published 1865.

Jorgenson, Andrew. 2014. "Economic Development and the Carbon Intensity of Human Well-Being." *Nature Climate Change* 4: 186–189.

Jorgenson, Andrew, and Brett Clark. 2011. "Societies Consuming Nature: A Panel Study of the Ecological Footprints of Nations, 1960–2003." *Social Science Research* 40: 226–244.

Jorgenson, Andrew, and Brett Clark. 2012. "Are the Economy and the Environment Decoupling?: A Comparative International Study, 1960–2005." *American Journal of Sociology* 118: 1–44.

Jorgenson, Andrew, Brett Clark, and Jeffrey Kentor. 2010. "Militarization and the Environment: A Panel Study of Carbon Dioxide Emissions and the Ecological Footprints of Nations, 1970–2000." *Global Environmental Politics* 10: 7–29.

Jorgenson, Andrew, Christopher Dick, and Matthew Mahutga. 2007. "Foreign Investment Dependence and the Environment: An Ecostructural Approach." *Social Problems* 54: 371–394.

Jorgenson, Andrew, Christopher Dick, and John Shandra. 2011. "World Economy, World Society, and Environmental Harms in Less-Developed Countries." *Sociological Inquiry* 81: 53–87.

Jorgenson, Andrew, Shirley Fiske, Klaus Hubacek, Jia Li, Tom McGovern, Torben Rick, Juliet Schor, William Solecki, Richard York, and Ariela Zycherman. 2019. "Social Science Perspectives on Drivers of and Responses to Global Climate Change." *Wiley Interdisciplinary Reviews: Climate Change* 10, no. 1: e554. https://doi.org/10.1002/wcc.554.

Jorgenson, Andrew, Terrence Hill, Brett Clark, Ryan Thombs, Peter Ore, Kelly Balistreri, and Jennifer Givens. 2020. "Power, Proximity, and Physiology: Does Income Inequality and Racial Composition Amplify the Impacts of Air Pollution on Life Expectancy in the United States?" *Environmental Research Letters*. https://doi.org/10.1088/1748-9326/ab6789.

Joskow, Paul, and Richard Schmalensee. 1987. "The Performance of Coal-Burning Electric Generating Units in the United States: 1960–1980." *Journal of Applied Econometrics* 2: 85–109.

Kahneman, Daniel. 2011. *Thinking, Fast and Slow.* New York: Farrar, Straus and Giroux.

Kaufman, Noah, and Kate Gordon. 2018. *The Energy, Economic, and Emissions Impacts of a Federal US Carbon Tax*. New York: Columbia Center on Global Energy Policy.

Kennedy, Emily Huddart, Harvey Krahn, and Naomi Krogman. 2014. "Egregious Emitters: Disproportionality in Household Carbon Footprints." *Environment and Behavior* 46, no. 5: 535–555.

Kenner, Dario. 2019. *Carbon Inequality: The Role of the Richest in Climate Change*. Abingdon, UK: Routledge.

Keohane, Robert O. 2014. "The Global Politics of Climate Change: Challenge for Political Science." 2014 James Madison Lecture. *PS: Political Science and Politics* 48, no. 1: 19–26. https://doi.org/10.1017/S1049096514001541.

Kessel, Jonah, and Hiroko Tabuchi. 2019. "It's a Vast, Invisible Climate Menace. We Made It Visible." *New York Times*. December 12.

Keyes, Amelia, Kathleen Lambert, Dallas Burtraw, Jonathan Buonocore, Jonathan Levy, and Charles Driscoll. 2019. "The Affordable Clean Energy Rule and the Impact of Emissions Rebound on Carbon Dioxide and Criteria Air Pollutant Emissions." *Environmental Research Letters* 14: 044018.

Khazzoom, J. Daniel. 1980. "Economic Implications of Mandated Efficiency Standards for Household Appliances." *Energy Journal* 1, no. 4: 85–89.

Khazzoom, J. Daniel. 1982. "Response to Besen and Johnson's Comment on Economic Implications of Mandated Efficiency Standards for Household Appliances." *Energy Journal* 3, no. 1: 117–124.

Kirk, Geoffrey, ed. 1982. *Schumacher on Energy*. London: Jonathan Cape.

Kirsch, Jan. 2019. *Banking on Climate Change: Fossil Fuel Finance Report Card 2019*. San Francisco: Rainforest Action Network.

Krugman, Paul. 2014. "The Big Green Test: Conservatives and Climate Change." *New York Times*. June 23.

Le Quiere, Corrine, Robbie M. Andrew, Pierre Friedlingstein, Stephen Sitch, Judith Hauck, Julia Pongratz, Penelope A. Pickers et al. 2018. "Global Carbon Budget 2018." *Earth Systems Science Data* 10: 2141–2194.

Lewis, Steven, and Anna Hogan. 2019. "Reform First and Ask Questions Later?: The Implications of (Fast) Schooling Policy and 'Silver Bullet' Solutions." *Critical Studies in Education* 60, no. 1: 1–18.

Lind, Benjamin, and Judith Stepan-Norris. 2011. "The Relationality of Movements: Movement and Countermovement Resources, Infrastructure, and Leadership in the Los Angeles Tenants' Rights Mobilization, 1976–1979." *American Journal of Sociology* 116, no. 5: 1564–1609.

Liu, Jianguo, Thomas Dietz, Stephen R. Carpenter, Marina Alberti, Carl Folke, Emilio Moran, Alice N. Pell et al. 2007. "Complexity of Coupled Human and Natural Systems." *Science* 317: 1513–1516.

Liu, Jianguo, Vanessa Hull, Mateus Batistella, Ruth DeFries, Thomas Dietz, Feng Fu, Thomas W. Hertel et al. 2013. "Framing Sustainability in a Telecoupled World." *Ecology and Society* 18, no. 2: 26. http://dx.doi.org/10.5751/ES-05873-180226.

Longest, Kyle, and Stephen Vaisey. 2008. "Fuzzy: A Program for Performing Qualitative Comparative Analyses (QCA) in Stata." *Stata Journal* 8: 79–104.

Longhofer, Wesley, and Andrew Jorgenson. 2017. "Decoupling Reconsidered: Does World Society Integration Influence the Relationship Between the Environment and Economic Development?" *Social Science Research* 65: 17–29.

Longhofer, Wesley, and Evan Schofer. 2010. "National and Global Origins of Environmental Association." *American Sociological Review* 75: 505–533.

Longhofer, Wesley, Evan Schofer, Natasha Miric, and David John Frank. 2016. "NGOs, INGOs, and Social Change: Environmental Policy Reform in the Developing World, 1970–2010." *Social Forces* 94: 1743–1768.

Luders, Joseph. 2006. "The Economics of Movement Success: Business Responses to Civil Rights Mobilization." *American Journal of Sociology* 111, no. 4: 963–998.

Lutsey, Nicholas, and Daniel Sperling. 2008. "America's Bottom-Up Climate Change Mitigation Strategy." *Energy Policy* 36: 673–685.

Lyon, Thomas P., and Yin Haitao. 2010. "Why Do States Adopt Renewable Portfolio Standards? An Empirical Investigation." *Energy Journal* 31, no. 3: 133–157.

Lyon, Thomas P., and John W. Maxwell. 2011. "Greenwash: Corporate Environmental Disclosure Under Threat of Audit." *Journal of Economics and Management Strategy* 20: 3–41.

Manzo, Gianluca. 2014. "Potentialities and Limitations of Agent-Based Simulations." *Review of French Sociology* 55: 653–688.

Marland, G., T. A. Boden, and R. J. Andres. 2008. "Global, Regional, and National Fossil-Fuel CO_2 Emissions." In "Trends: A Compendium of Data on Global Change." Carbon Dioxide Information Analysis Center, Oak Ridge National Laboratory, U.S. Department of Energy, Oak Ridge, TN. https://cdiac.ess-dive.lbl.gov/trends/emis/overview.html.

Marquis, Christopher, Michael W. Toffel, and Yanhua Zhou. 2016. "Scrutiny, Norms, and Selective Disclosure: A Global Study of Greenwashing." *Organization Science* 27: 483–504.

Martin, Geoff, and Eri Saikawa. 2017. "Effectiveness of State Climate and Energy Policies in Reducing Power-Sector CO_2 Emissions." *Nature Climate Change* 7: 912–919.

Marx, Karl, and Friedrich Engels. 1888 [1848]. *The Manifesto of the Communist Party*. Moscow: Foreign Languages.

Matisoff, Daniel. 2008. "The Adoption of State Climate Change Policies and Renewable Portfolio Standards: Regional Diffusion or Internal Determinants?" *Review of Policy Research* 25: 527–546.

McAdam, Doug. 2017. "Social Movement Theory and the Prospects for Climate Change Activism in the United States." *Annual Review of Political Science* 20: 189–208.

McAdam, Doug, and Hilary Boudet. 2012. *Putting Social Movements in Their Place: Opposition to Energy Projects, 2000–2005*. New York: Cambridge University Press.

McAdam, Doug, Hilary Schaffer Boudet, Jennifer Davis, Ryan Orr, W. Richard Scott, and Raymond Levitt. 2010. " 'Site Fights': Explaining Opposition to Pipeline Projects in the Developing World." *Sociological Forum* 25, no. 3: 401–427.

McCright, Aaron, and Riley Dunlap. 2010. "Anti-reflexivity: The American Conservative Movement's Success in Undermining Climate Science and Policy." *Theory, Culture and Society* 27, no. 2–3: 100–133.

McKinsey and Company. 2009. *Unlocking Energy Efficiency in the U.S. Economy*. New York: McKinsey. https://www.mckinsey.com/~/media/mckinsey/dotcom/client_service/epng/pdfs/unlocking%20energy%20efficiency/us_energy_efficiency_exc_summary.ashx.

McLaughlin, Paul. 2011. "Climate Change, Adaptation, and Vulnerability: Reconceptualizing Societal-Environment Interaction Within a Socially Constructed Adaptive Landscape." *Organization and Environment* 24: 269–291.

McNeill, John R. 2000. *Something New Under the Sun: An Environmental History of the Twentieth-Century World*. New York: Norton.

Mendel, Jerry M., and Charles C. Ragin. 2011. "fsQCA: Dialog Between Jerry M. Mendel and Charles C. Ragin." USC-SIPI Report 411. University of Southern California, Los Angeles. http://sipi.usc.edu/reports/pdfs/Originals/USC-SIPI-411.pdf.

Meyer, John W., Francisco Ramirez, and Yasemin Soysal. 1992. "World Expansion of Mass Education, 1870–1980." *Sociology of Education* 65, no. 2: 128–149.

Meyer, John W., and Brian Rowan. 1977. "Institutionalized Organizations: Formal Structure as Ceremony and Myth." *American Journal of Sociology* 83, no. 2: 340–363.

Michaels, Robert. 2008. "National Renewable Portfolio Standard: Smart Policy or Misguided Gesture?" *Energy Law Journal* 29: 79–119.

Miller, Danny. 1987. "The Genesis of Configuration." *Academy of Management Review* 12: 686–701.

Millner, Antony, and Thomas McDermott. 2016. "Model Confirmation in Climate Economics." *Proceedings of the National Academy of Sciences* 113, no. 31: 8675–8680.

Minnesma, Marjan. 2019. "Not Slashing Emissions? See You in Court." *Nature* 576: 379–382.

Mohai, Paul, David Pellow, and Timmons Roberts. 2009. "Environmental Justice." *Annual Review of Environmental Resources* 34: 404–430.

Mol, Arthur. 2001. *Globalization and Environmental Reform*. Cambridge, MA: MIT Press.

Mol, Arthur P. J., Gert Spaargaren, and David A. Sonnenfeld. 2014. "Ecological Modernization Theory: Taking Stock, Moving Forward." In *The Routledge International Handbook of Social and Environmental Change*, ed. Stewart Lockie, David Sonnenfeld, and Dana R. Fisher, 15–30. New York: Routledge.

Morgan Energy Solutions. 2013. "Taichung Power Plant: World's Largest Coal Fired Plant." January 16.

Nagel, Joane, Jeffrey Broadbent, and Thomas Dietz. 2010. *Workshop on Sociological Perspectives on Climate Change*. Washington, DC: American Sociological Association.

National Energy Technology Laboratory. 2008. *Reducing CO_2 Emissions by Improving the Efficiency of the Existing Coal-Fired Power Plant Fleet*. www.netl.doe.gov/energy-analyses/pubs/CFPP%20Efficiency-FINAL.pdf.

National Renewable Energy Laboratory (NREL). 2010. *Evaluating Renewable Portfolio Standards and Carbon Cap Scenarios in the U.S. Electric Sector*. NREL/TP-6A2-48258. Golden, CO: NREL.

National Research Council. 1983. *Changing Climate: Report of the Carbon Dioxide Assessment Committee*. Washington, DC: National Academies Press.

Nemet, Gregory F., and Erin Baker. 2009. "Demand Subsidies Versus R&D: Comparing the Uncertain Impacts of Policy on a Pre-commercial Low-Carbon Energy Technology." *Energy Journal* 30: 49–80.

Nelson, Richard R., and Sidney G. Winter. 1982. *An Evolutionary Theory of Economic Change*. Cambridge, MA: Belknap Press.

Nilsson, Andreas, Magnus Bergquist, and Wesley P. Schultz. 2017. "Spillover Effects in Environmental Behaviors, Across Time and Context: A Review and Research Agenda." *Environmental Education Research* 23, no. 4: 573–589. https://doi.org/10.1080/13504622.2016.1250148DOI: 10.1080/13504622.2016.1250148.

Oxfam. 2015. *Extreme Carbon Inequality*. Media Briefing. Nairobi: Oxfam, December 2. https://www-cdn.oxfam.org/s3fs-public/file_attachments/mb-extreme-carbon-inequality-021215-en.pdf.

Palmer, Karen, and Anthony Paul. 2015. *A Primer on Comprehensive Policy Options for States to Comply with the Clean Power Plan*. Washington, DC: Resources for the Future.

Peck, Jamie, and Nik Theodore. 2015. *Fast Policy: Experimental Statecraft at the Thresholds of Neoliberalism*. Minneapolis: University of Minnesota Press.

Pellow, David N. 2000. "Environmental Inequality Formation: Toward a Theory of Environmental Justice." *American Behavioral Scientist* 43: 581–601.

Pellow, David. 2002. *Garbage Wars: The Struggle for Environmental Justice in Chicago*. Cambridge, MA: MIT Press.

Pellow, David, and Hollie Brehm. 2013. "An Environmental Sociology for the Twenty-First Century." *Annual Review of Sociology* 39: 229–250.

Perrow, Charles. 2010. "Organizations and Global Warming," In *Routledge Handbook of Climate Change and Society*, ed. Constance Lever-Tracy, 59–77. New York: Routledge.

Perrow, Charles, and Simone Pulver. 2015. "Organizations and Markets." In *Climate Change and Society: Sociological Perspectives*, ed. R. Dunlap and R. Brulle. New York: Oxford University Press.

Peters, Glen, Jan Minx, Christopher Weber, and Ottmar Edenhofer. 2011. "Growth in Emission Transfers Via International Trade from 1990 to 2008." *Proceedings of the National Academy of Sciences* 108, no. 21: 8903–8908.

Pew Center on Global Climate Change. 2006. *Climate Change 101: State Action*. Arlington, VA: Pew.

Podobnik, Bruce. 2006. *Global Energy Shifts: Fostering Sustainability in a Turbulent Age*. Philadelphia: Temple University Press.

Polimeni, John, Kozo Mayumi, Mario Giampietro, and Blake Alcot. 2008. *The Jevons Paradox and the Myth of Resource Efficiency Improvements*. London: Earthscan.

Political Economy Research Institute. 2018. "Toxic 100 Air Polluter Index (2018 Report Based on 2015 Data)." University of Massachusetts, Amherst. https://www.peri.umass.edu/toxic-100-air-polluters-index-2018-report-based-on-2015-data.

Pollin, Robert. 2019. "Advancing a Viable Global Climate Stabilization Project: Degrowth Versus the Green New Deal." *Review of Radical Political Economics* 51: 311–319.

Prasad, Monica. 2018. "Problem-Solving Sociology." *Contemporary Sociology* 47: 393–398.

Prasad, Monica, and Steven Munch. 2012. "State-Level Renewable Electricity Policies and Reductions in Carbon Emissions." *Energy Policy* 45: 237–242.

Prechel, Harland. 2012. "Corporate Power and U.S. Economic and Environmental Policy, 1978–2008." *Cambridge Journal of Regions, Economy and Society* 5: 357–375.

Prechel, Harland, and Alesha Istvan. 2016. "Disproportionality of Corporations' Environmental Pollution in the Electrical Energy Industry." *Sociological Perspectives* 59, no. 3: 505–527.

Prechel, Harland, and George Touche. 2014. "The Effects of Organizational Characteristics and State Environmental Policies on Sulfur-Dioxide Pollution in U.S. Electrical Energy Corporations." *Social Science Quarterly* 95: 76–96.

Prechel, Harland, and Lu Zheng. 2012. "Corporate Characteristics, Political Embeddedness, and Environmental Pollution by Largest U.S. Corporations." *Social Forces* 90: 947–970.

Pulver, Simone. 2011. "Corporate Responses." In *The Oxford Handbook of Climate Change and Society*, ed. J. Dryzek, R. Norgaard, and D. Schlosberg, 581–592. Oxford: Oxford University Press.

Putnam, Robert. 2000. *Bowling Alone: The Collapse and Revival of American Community*. New York: Simon & Schuster.

Rabe, Barry. 2007. "Race to the Top: The Expanding Role of U.S. State Renewable Portfolio Standards." *Sustainable Development Law and Policy* 7: 10–16.

Rabe, Barry. 2008. "States on Steroids: The Intergovernmental Odyssey of American Climate Policy." *Review of Policy Research* 25: 105–128.

Rabe, Barry. 2013. "Racing to the Top, the Bottom, or the Middle of the Pack? The Evolving State Government Role in Environmental Protection." In *Environmental Policy: New Directions for the Twenty-First Century*, ed. N. Vig and M. Kraft, 30–53. Washington, DC: CQ Press.

Rabe-Hesketh, Sophia, and Anders Skrondal. 2008. *Multilevel and Longitudinal Modeling Using Stata*. 2nd ed. College Station, TX: Stata Press.

Ragin, Charles C. 2000. *Fuzzy-Set Social Science*. Chicago: University of Chicago Press.

Ragin, Charles C., and Peer C. Fiss. 2008. "Net Effects Versus Configurations: An Empirical Demonstration." In *Redesigning Social Inquiry: Fuzzy Sets and Beyond*, ed. Charles C. Ragin, 190–212. Chicago: University of Chicago Press.

Ramirez, Francisco, Yasemin Soysal, and Suzanne Shanahan. 1997. "The Changing Logic of Political Citizenship: Cross-National Acquisition of Women's Suffrage Rights, 1890 to 1990." *American Sociological Review* 62, no. 5: 735–745.

Rayner, Steve. 2010. "How to Eat an Elephant: A Bottom-Up Approach to Climate Policy." *Climate Policy* 10: 615–621.

Reuters. 2014. "Russian Firm Studying World's Largest Coal-Fired Plant to Supply China." May 26. https://www.reuters.com/article/russia-interrao-plant/russian-firm-studying-worlds-largest-coal-fired-plant-to-supply-china-idUSL6N0OC30R20140526.

Revesz, Richard, Peter Howard, Kenneth Arrow, Lawrence Goulder, Robert Kopp, Michael Livermore, Michael Oppenheimer, and Thomas Sterner. 2014. "Global Warming: Improve Economic Models of Climate Change." *Nature* 508: 173–175.

Rihoux, Benoit, and Charles Ragin, eds. 2009. *Configurational Comparative Methods: Qualitative Comparative Analysis (QCA) and Related Techniques*. Thousand Oaks, CA: Sage.

Roberts, David. 2019. "A Major Utility Is Moving Toward 100 Percent Clean Energy Faster than Expected." *Vox*. May 29. https://www.vox.com/energy-and-environment/2018/12/5/18126920/xcel-energy-100-percent-clean-carbon-free.

Roberts, Timmons, Peter Grimes, and Jodie Manale. 2003. "Social Roots of Global Environmental Change: A World-Systems Analysis of Carbon Dioxide Emissions." *Journal of World-Systems Research* 9, no. 2: 277–315.

Roberts, Timmons, and Bradley Parks. 2007. *A Climate of Injustice: Global Inequality, North-South Politics, and Climate Policy*. Cambridge, MA: MIT Press.

Robertson, Anya, and Mary Collins. 2019. "Super Emitters in the United States Coal-Fired Electric Utility Industry: Comparing Disproportionate Emissions Across Facilities and Parent Companies." *Environmental Sociology* 5: 70–81.

Rosa, Eugene, and Thomas Dietz. 2012. "Human Drivers of National Greenhouse Gas Emissions." *Nature Climate Change* 2: 581–586.

Rosa, Eugene, and Lauren Richter. "Durkheim on the Environment: Ex Libris or Ex Cathedra? Introduction to Inaugural Lecture to a Course in Social Science, 1887–1888." *Organization and Environment* 21, no. 2 (2008): 182–187.

Roscigno, Vincent, and Randy Hodson. 2004. "The Organizational and Social Foundations of Worker Resistance." *American Sociological Review* 69: 14–39.

Rubinson, Claude. 2019. "Presenting Qualitative Comparative Analysis: Notation, Tabular Layout, and Visualization." *Methodological Innovations* 12 (May–August): 1–22.

Rudel, Thomas K. 2009. "How Do People Transform Landscapes?: A Sociological Perspective on Suburban Sprawl and Tropical Deforestation." *American Journal of Sociology* 115: 129–154.

Sabel, Charles, and David G. Victor. 2017. "Governing Global Problems Under Uncertainty: Making Bottom-Up Climate Policy Work." *Climatic Change* 44, no. 1: 15–27.

Sabel, Charles, and Jonathan Zeitlin. 2012. "Experimentalist Governance." In *The Oxford Handbook of Governance*, ed. D. Levi-Faur. New York: Oxford University Press.

Schelly, David, and Paul Stretesky. 2009. "An Analysis of the 'Path of Least Resistance' Argument in Three Environmental Justice Success Cases." *Society and Natural Resources* 22: 369–380.

Schneider, Carsten Q., and Claudius Wagemann. 2006. "Reducing Complexity in Qualitative Comparative Analysis (QCA): Remote and Proximate Factors and the Consolidation of Democracy." *European Journal of Political Research* 45, no. 5: 751–786.

Schofer, Evan, and Ann Hironaka. 2005. "The Effects of World Society on Environmental Protection Outcomes." *Social Forces* 84: 25–47.

Schor, Juliet, and Andrew Jorgenson. 2019. "Is It Too Late for Growth?" *Review of Radical Political Economics* 51: 320–329.

Schurman, Rachel, and William Munro. 2009. "Targeting Capital: A Cultural Economy Approach to Understanding the Efficacy of Two Anti-genetic Engineering Movements." *American Journal of Sociology* 115, no. 1: 155–202.

Scott, W. Richard. 1992. *Organizations: Rational, Natural and Open Systems*. 3rd ed. Englewood Cliffs, NJ: Prentice-Hall.

Scott, W. Richard, and Gerald Davis. 2016. *Organizations and Organizing: Rational, Natural, and Open Perspectives*. New York: Routledge.

Selin, Henrik, and Stacey D. VanDeveer. 2009. "Climate Leadership in Northeast North America." In *Changing Climates in North American Politics: Institutions, Policymaking, and Multilevel Governance*, ed. H. Selin and S. VanDeveer, 111–136. Cambridge, MA: MIT Press).

Shandra, John, Christopher Leckband, Laura McKinney, and Bruce London. 2009. "Ecologically Unequal Exchange, World Polity, and Biodiversity Loss: A Cross-National Analysis of Threatened Mammals." *International Journal of Comparative Sociology* 50: 285–310.

Shandra, John, Bruce London, Owen Wooley, and John Williamson. 2004. "International Nongovernmental Organizations and Carbon Dioxide Emissions in the Developing World: A Quantitative, Cross-National Analysis." *Sociological Inquiry* 74: 520–544.

Shiller, Robert. 2012. *Finance and the Good Society*. Princeton, NJ: Princeton University Press.

Shorette, Kristen. 2012. "Outcomes of Global Environmentalism: Longitudinal and Cross-National Trends in Chemical Fertilizer and Pesticide Use." *Social Forces* 91: 299–325.

Shorette, Kristen, Kent Henderson, Jamie Sommer, and Wesley Longhofer. 2017. "World Society and the Natural Environment." *Sociology Compass* 11, no. 10. https://doi.org/10.1111/soc4.12511.

Shriver, Thomas, Alison Adams, and Sherry Cable. 2013. "Discursive Obstruction and Elite Opposition to Environmental Activism in the Czech Republic." *Social Forces* 91: 873–893.

Shriver, Thomas, Alison Adams, and Chris Messer. 2014. "Power, Quiescence, and Pollution: The Suppression of Environmental Grievances." *Social Currents* 1: 275–292.

Shwom, Rachael. 2009. "Strengthening Sociological Perspectives on Organizations and the Environment." *Organization and Environment* 22: 271–292.

Shwom, Rachael. 2011. "A Middle Range Theorization of Energy Politics: The Struggle for Energy Efficient Appliances." *Environmental Politics* 20: 705–726.

Sine, Wesley, and Brandon Lee. 2009. "Tilting at Windmills?: The Environmental Movement and the Emergence of the U.S. Wind Energy Sector." *Administrative Science Quarterly* 54: 123–155.

Skocpol, Theda. 2014. "How the Scholars Strategy Network Helps Academics Gain Public Influence." *Perspectives on Politics* 12, no. 3: 695–703.

Sorrell, Steve. 2009. "Jevons' Paradox Revisited: The Evidence for Backfire from Improved Energy Efficiency." *Energy Policy* 37: 1456–1469.

Sovacool, Benjamin. 2014. "What Are We Doing Here?: Analyzing Fifteen Years of Energy Scholarship and Proposing a Social Science Research Agenda." *Energy Research and Social Science* 1: 1–29.

Sovacool, Benjamin, Jonn Axsen, and Steve Sorrell. 2018. "Promoting Novelty, Rigor, and Style in Energy Social Science: Towards Codes of Practice for Appropriate Methods and Research Design." *Energy Research and Social Science* 45: 12–42.

Stall, Susan, and Randy Stoecker. 1998. "Community Organizing or Organizing Community?" *Gender and Society* 12, no. 6: 729–756.

Steffen, Will, Johan Rockström, Katherine Richardson, Timothy M. Lenton, Carl Folke, Diana Liverman, Colin P. Summerhayes et al. 2018. "Trajectories of the Earth System in the Anthropocene." *Proceedings of the National Academy of Sciences* 115, no. 33: 8252–8259.

Stern, Nicholas. 2016. "Current Climate Models Are Grossly Misleading." *Nature* 530: 407–409.

Taipei Times. 2008. "Taichung Power Plant World's Worst Polluter: Survey." September 4, 11.

Taiwan News. 2019. "Central Taiwan Power Plant Fined Again for Exceeding Coal Use." December 14.

Thaler, Richard H. 2015. *Misbehaving: The Making of Behavioral Economics*. New York: Norton.

Thomas, Kimberly, Dean Hardy, Heather Lazrus, Michael Mendez, Ben Orlove, Isabel Rivera-Collazo, Timmons Roberts, Marcy Rockman, Benjamin Warner, and Robert Winthrop. 2019. "Explaining Differential Vulnerability to Climate Change: A Social Science Review." *Wiley Interdisciplinary Reviews: Climate Change* 10, no. 2: e565. https://doi.org/10.1002/wcc.565.

Thombs, Ryan. 2018. "The Transnational Tilt of the Treadmill and the Role of Trade Openness on Carbon Emissions: A Comparative International Study, 1965–2010." *Sociological Forum* 33: 422–442.

Thombs, Ryan, and Xiaorui Huang. 2019. "Uneven Decoupling: The Economic Growth–CO_2 Emissions Relationship in the Global North, 1870–2014." *Sociology of Development* 5: 410–427.

Thombs, Ryan, and Andrew Jorgenson. 2020. "The Political Economy of Renewable Portfolio Standards in the United States." *Energy Research and Social Science* 62: 101379.

Tolbert, Charles, Thomas Lyson, and Michael Irwin. 1998. "Local Capitalism, Civic Engagement, and Socioeconomic Well-Being." *Social Forces* 77: 401–428.

Tong, Dan, Qiang Zhang, Steve J. Davis, Fei Liu, Bo Zheng, Guannan Geng, Tao Xue et al. 2018. "Targeted Emission Reductions from Global Super-Polluting Power Plant Units." *Nature Sustainability* 1: 59–68.

Truelove, Heather Barnes, Amanda Carrico, Elke Weber, Kaitlin Toner Raimi, and Michael Vandenbergh. 2014. "Positive and Negative Spillover of Pro-environmental Behavior: An Integrative Review and Theoretical Framework." *Global Environmental Change* 29: 127–138.

United Nations Framework Convention on Climate Change. 2007. "Bali Action Plan." In "Report of the Conference of the Parties on Its Thirteenth Session." Bonn. https://unfccc.int/resource/docs/2007/cop13/eng/06a01.pdf.

Urry, John. 2014. "The Problem of Energy." *Theory, Culture and Society* 31, no. 5: 3–20.

U.S. Energy Information Administration. 2011. *U.S. Energy-Related Carbon Emissions, 2010*. Washington, DC: U.S. Department of Energy.

U.S. Environmental Protection Agency (EPA). 2009. *Energy Efficiency as a Low-Cost Resource for Achieving Carbon Emissions Reductions*. Washington, DC: EPA.

U.S. Environmental Protection Agency (EPA). 2010. *Available and Emerging Technologies for Reducing Greenhouse Gas Emissions from Coal-Fired Electric Generating Units*. Washington, DC: EPA.

Vasi, Ion Bogdan. 2009. "Social Movements and Industry Development: The Environmental Movement's Impact on the Wind Energy Industry." *Mobilization* 14, no. 3: 315–336.

Vasi, Ion Bogdan. 2011. *Winds of Change. The Environmental Movement and the Global Development of the Wind Energy Industry*. Oxford: Oxford University Press.

Vasi, Ion Bogdan, and Brayden King. 2012. "Social Movements, Risk Perceptions, and Economic Outcomes: The Effect of Primary and Secondary Stakeholder Activism on Firms' Perceived Environmental Risk and Financial Performance." *American Sociological Review* 77, no. 4: 573–597.

Victor, David. 2011. *Global Warming Gridlock: Creating More Effective Strategies for Protecting the Planet*. New York: Cambridge University Press.

Victor, David G., Joshua C. House, and Sarah Joy. 2005. "A Madisonian Approach to Climate Policy." *Science* 309: 1820–1821.

Wackernagel, Matthis, and William Rees. 1998. *Our Ecological Footprint: Reducing Human Impact on the Earth*. Gabriola Island, BC: New Society Publishers.

Walsh, Edward, Rex Warland, and D. Clayton Smith. 1993. "Backyards, NIMBYs, and Incinerator Sitings: Implications for Social Movement Theory." *Social Problems* 40, no. 1: 25–38.

Weber, Max. 1978. *Economy and Society: An Outline of Interpretive Sociology*. Berkeley: University of California Press. First published 1922.

Weber, Max. 1992. *The Protestant Ethic and the Spirit of Capitalism*, trans. T. Parsons. London: Routledge. First published 1904–1905.

Weitzman, Martin L. 2011. "Fat-Tailed Uncertainty in the Economics of Catastrophic Climate Change." *Review of Environmental Economics and Policy* 5, no. 2: 275–292.

Weitzman, Martin L. 2015. "A Review of William Nordhaus' *The Climate Casino: Risk, Uncertainty, and Economics for a Warming World*." *Review of Environmental Economics and Policy* 9, no. 1: 145–156.

Westphal, James D., and Edward J. Zajac. 2001. "Decoupling Policy from Practice: The Case of Stock Repurchase Programs." *Administrative Science Quarterly* 46: 202–228.

Wheeler, David, and Kevin Ummel. 2008. "Calculating CARMA: Global Estimation of CO_2 Emissions from the Power Sector." Working Paper 145, Center for Global Development, Washington, DC. https://www.cgdev.org/publication/calculating-carma-global-estimation-co2-emissions-power-sector-working-paper-145.

World Resources Institute (WRI). 2006. *Target: Intensity, an Analysis of Greenhouse Gas Intensity Targets*. Washington, DC: WRI.

Yale School of Forestry and Environmental Studies. 2017. "Poll: Majority in All States, Congressional Districts Support Clean Power Plan." https://environment.yale.edu/news/article/poll-majority-support-for-clean-power-plan-in-all-states-congressional-districts/.

York, Richard. 2006. "Ecological Paradoxes: William Stanley Jevons and the Paperless Office." *Human Ecology Review* 13, no. 2: 143–147.

York, Richard. 2010. "The Paradox at the Heart of Modernity: The Carbon Efficiency of the Global Economy." *International Journal of Sociology* 40, no. 2: 6–22.

York, Richard, Christina Ergas, Eugene Rosa, and Thomas Dietz. 2011. "It's a Material World: Trends in Material Extraction in China, India, Indonesia, and Japan." *Nature and Culture* 6, no. 2: 103–122.

York, Richard, and Julius McGee. 2016. "Understanding the Jevons Paradox." *Environmental Sociology* 2: 77–87.

York, Richard, Eugene Rosa, and Thomas Dietz. 2003. "Footprints on the Earth: The Environmental Consequences of Modernity." *American Sociological Review* 68: 279–300.

Zeitlin, Jonathan. 2016. "EU Experimentalist Governance in Times of Crisis." *West European Politics* 39, no. 5: 1073–1094.

Zipf, George Kingsley. *Human Behavior and the Principle of Least Effort: An Introduction to Human Ecology*. Cambridge, MA: Addison-Wesley, 1949; Eastford, CT: Martino, 2012.

INDEX

Page numbers in *italics* indicate figures or tables.